Graphing Calculator Manual

Judith A. Penna
Indiana University Purdue University Indianapolis

Elementary and Intermediate Algebra Graphs and Models
Second Edition

Marvin L. Bittinger
Indiana University Purdue University Indianapolis

David J. Ellenbogen
Community College of Vermont

Barbara L. Johnson
Indiana University Purdue University Indianapolis

PEARSON

Addison Wesley

Boston   San Francisco   New York
London   Toronto   Sydney   Tokyo   Singapore   Madrid
Mexico City   Munich   Paris   Cape Town   Hong Kong   Montreal

Reproduced by Pearson Addison-Wesley from electronic files supplied by the author.

Copyright © 2004 Pearson Education, Inc.
Publishing as Pearson Addison-Wesley, 75 Arlington Street, Boston, MA 02116

All rights reserved. No part of this publication may be reproduced, stored in a retrieval system, or transmitted, in any form or by any means, electronic, mechanical, photocopying, recording, or otherwise, without the prior written permission of the publisher. Printed in the United States of America.

ISBN   0-321-16868-2

5 6 VHG 06 05

Contents

The TI-83 and TI-83 Plus Graphics Calculators 1

The TI-86 Graphics Calculator 61

The TI-89 Graphics Calculator 117

TI-83 and TI-83 Plus Index I-1

TI-86 Index I-5

TI-89 Index I-9

The TI-83 and TI-83 Plus Graphics Calculators

Chapter 1
Introduction to Algebraic Expressions

GETTING STARTED

Press ON to turn on the TI-83 or TI-83 Plus graphing calculator. (ON is the key at the bottom left-hand corner of the keypad.) You should see a blinking rectangle, or cursor, on the screen. If you do not see the cursor, try adjusting the display contrast. To do this, first press 2nd . (2nd is the yellow key in the left column of the keypad.) Then press and hold △ to increase the contrast or ▽ to decrease the contrast.

To turn the calculator off, press 2nd OFF . (OFF is the second operation associated with the ON key.) The calculator will turn itself off automatically after about five minutes without any activity.

Press MODE to display the MODE settings. Initially you should select the settings on the left side of the display.

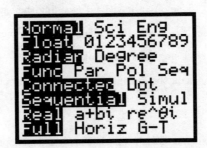

To change a setting on the Mode screen use ▽ or △ to move the blinking cursor to the line of that setting. Then use ▷ or ◁ to move the cursor to the desired setting and press ENTER . Press CLEAR or 2nd QUIT to leave the MODE screen. (QUIT is the second operation associated with the MODE key.) Pressing CLEAR or 2nd QUIT will take you to the home screen where computations are performed.

The TI-83 and TI-83 Plus graphing calculators are very similar in many respects. For that reason, most of the keystrokes and instructions presented in this section of the graphing calculator manual will apply to both calculators. Where they differ, keystrokes and instructions for using the TI-83 will be given first, followed by those for the TI-83 Plus.

It will be helpful to read the Introduction to the Graphing Calculator on pages 4 and 5 of your textbook as well as the Getting Started section of your graphing calculator Guidebook before proceeding.

EVALUATING EXPRESSIONS

To evaluate expressions we substitute values for the variables.

Section 1.1, Example 4 Use a graphing calculator to evaluate $3xy + x$ for $x = 65$ and $y = 92$.

Enter the expression in the calculator, replacing x with 65 and y with 92. Press 3 × 6 5 × 9 2 + 6 5 ENTER . The calculator returns the value of the expression, 18,005.

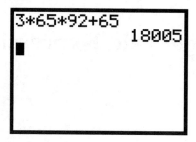

You can recall and edit your entry if necessary. If, for instance, in the expression above you pressed 8 instead of 9, first press 2nd ENTRY to return to the last entry. (ENTRY is the second operation associated with the ENTER key.) Then use the ◁ key to move the cursor to 8 and press 9 to overwrite it. If you forgot to type the 2, move the cursor to the plus sign; then press 2nd INS 2 to insert the 2 before the plus sign. (INS is the second operation associated with the DEL key.) You can continue to insert symbols immediately after the first insertion without pressing 2nd INS again. If you typed 31 instead of 3, move the cursor to 1 and press DEL. This will delete the 1. If you notice that an entry needs to be edited before you press ENTER to perform the computation, the editing can be done directly without recalling the entry.

The keystrokes 2nd ENTRY can be used repeatedly to recall entries preceding the last one. Pressing 2nd ENTRY twice, for example, will recall the next to last entry. Using these keystrokes a third time recalls the third to last entry and so on. The number of entries that can be recalled depends on the amount of storage they occupy in the calculator's memory.

USING A MENU

A menu is a list of options that appears when a key is pressed. Thus, multiple options, and sometimes multiple menus, may be accessed by pressing one key. For example, the following screen appears when the MATH key is pressed. We see the names of four submenus, MATH, NUM, CPX, and PRB as well as the options in the MATH submenu.

We can copy an item from a menu to the home screen either by using the up or down arrow key to highlight its number and then pressing ENTER or by simply pressing the number of the item. The down-arrow beside item 7 in the menu above indicates that there are additional items in the menu. Use the ▽ key to scroll down to them.

The next example involves choosing an option from a menu.

Section 1.3, Example 12 Use a graphing calculator to find fraction notation for $\frac{2}{15} + \frac{7}{12}$.

Enter $\frac{2}{15} + \frac{7}{12}$ by pressing 2 $\boxed{\div}$ 1 5 $\boxed{+}$ 7 $\boxed{\div}$ 1 2. Now press $\boxed{\text{ENTER}}$ to find decimal notation for the sum. To convert this to fraction notation we select the ▷Frac feature from the MATH submenu of the MATH menu. Press $\boxed{\text{MATH}}$ 1 $\boxed{\text{ENTER}}$ or $\boxed{\text{MATH}}$ $\boxed{\text{ENTER}}$ $\boxed{\text{ENTER}}$. These keystrokes recall the previous answer and then convert it to fraction notation. Note that this conversion must be done immediately after the calculation is performed in order to have the result of the calculation available for the conversion.

We can also find fraction notation for this sum in one step by selecting the ▷Frac operation before the sum is computed. When this is done, decimal notation does not appear on the screen. Press 2 $\boxed{\div}$ 1 5 $\boxed{+}$ 7 $\boxed{\div}$ 1 2 $\boxed{\text{MATH}}$ 1 $\boxed{\text{ENTER}}$ or 2 $\boxed{\div}$ 1 5 $\boxed{+}$ 7 $\boxed{\div}$ 1 2 $\boxed{\text{MATH}}$ $\boxed{\text{ENTER}}$ $\boxed{\text{ENTER}}$.

SQUARE ROOTS

Section 1.4, Example 5 Graph the real number $\sqrt{3}$ on a number line.

We can use the calculator to find a decimal approximation for $\sqrt{3}$. Press $\boxed{\text{2nd}}$ $\boxed{\sqrt{}}$ 3 $\boxed{)}$ $\boxed{\text{ENTER}}$. Note that the calculator supplies a left parenthesis along with the radical symbol. Although it is not necessary to supply a right parenthesis in this case, we will do so for completeness.

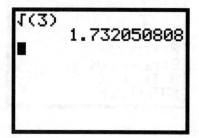

This approximation can now be used to locate $\sqrt{3}$ on the number line. The graph appears on page 33 of the text.

NEGATIVE NUMBERS AND ABSOLUTE VALUE

On the TI-83 and TI-83 Plus, $|x|$ is written abs(X). Absolute value notation is item 1 on the MATH NUM menu.

Section 1.4, Example 8 Find each absolute value: (a) $|-3|$; (b) $|7.2|$; (c) $|0|$.

When entering $|-3|$, keep in mind that the $\boxed{(-)}$ key in the bottom row of the keypad must be used to enter a negative number on the graphing calculator whereas the blue $\boxed{-}$ key in the right-hand column of the keypad is used to enter subtraction. Press $\boxed{\text{MATH}}$ $\boxed{\triangleright}$ $\boxed{\text{ENTER}}$ $\boxed{(-)}$ $\boxed{3}$ $\boxed{)}$ $\boxed{\text{ENTER}}$. To find $|7.2|$ use the keystrokes above, replacing -3 with 7 $\boxed{.}$ 2 and to find $|0|$ replace -3 with 0. Note that the calculator supplies a left parenthesis along with the absolute value notation. We supply the right parenthesis to close the absolute-value expression.

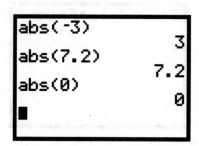

After finding $|-3|$, we could also find $|7.2|$ by recalling the previous entry ($|-3|$) and then editing that entry to replace -3 with 7.2. We could then recall the entry $|7.2|$ and edit it to find $|0|$. (See page 4 of this manual for a discussion of editing entries.)

Instead of pressing $\boxed{\text{MATH}}$ $\boxed{\triangleright}$ $\boxed{\text{ENTER}}$ to access "abs(" and copy it to the home screen, we could have pressed $\boxed{\text{MATH}}$ $\boxed{\triangleright}$ 1 since "abs(" is item 1 on the MATH NUM menu. Absolute value notation can also be found as the first item in the CATALOG

TI-83/83 Plus

and copied to the home screen. To do this press [2nd] [CATALOG] [ENTER]. (CATALOG is the second operation associated with the 0 numeric key.)

ORDER OF OPERATIONS; EXPONENTS AND GROUPING SYMBOLS

The TI-83 and TI-83 Plus follow the rules for order of operations.

Section 1.8, Example 8 Calculate: $\dfrac{12(9-7) + 4 \cdot 5}{2^4 + 3^2}$.

The fraction bar must be replaced with a set of parentheses around the entire numerator and another set of parentheses around the entire denominator when this expression is entered in the calculator. To enter an exponential expression, first enter the base, then use the [∧] key followed by the exponent. If the exponent is 2, the keystrokes [∧] 2 can be replaced with the single keystroke [x^2].

To enter the expression above and express the result in fraction notation, first press [(] 1 2 [(] 9 [−] 7 [)] [+] 4 [×] 5 [)] [÷] [(] 2 [∧] 4 [+] 3 [x^2] [)]. Remember to use the blue [−] key for subtraction rather than the [(−)] key, which is used to enter negative numbers. Then press [MATH] 1 or [MATH] [ENTER] to choose the Frac option from the MATH MATH menu. Finally, press [ENTER] to see the result.

Chapter 2
Equations, Inequalities, and Problem Solving

EDITING ENTRIES

In **Section 2.1** the procedure for editing entries is discussed on page 75. This procedure is described on page 4 of this manual.

EVALUATING FORMULAS; THE TABLE FEATURE: ASK MODE

Section 2.3, Example 2 Use the formula $B = 30a$, described in Example 1, to determine the minimum furnace output for well-insulated houses containing 800 ft^2, 1500 ft^2, 2400 ft^2, and 3600 ft^2.

First we replace B with y and a with x and enter the formula $y = 30x$ on the equation-editor screen as equation y_1. Press $\boxed{Y=}$ to access this screen. If any of Plot 1, Plot 2, and Plot 3 is turned on (highlighted), turn it off by using the arrow keys to move the blinking cursor over the plot name and pressing $\boxed{\text{ENTER}}$. If there is currently an expression displayed for y_1, clear it by positioning the cursor beside "$Y_1 =$" and pressing $\boxed{\text{CLEAR}}$. Do the same for expressions that appear on all other lines by using $\boxed{\triangledown}$ to move to a line and then pressing $\boxed{\text{CLEAR}}$. Then use $\boxed{\triangle}$ or $\boxed{\triangledown}$ to move the cursor to the top line beside "$Y_1 =$." Now press 3 0 $\boxed{X, T, \Theta, n}$ to enter the right-hand side of the equation on the "Y =" screen.

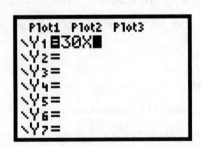

For an equation entered in the equation-editor screen, a table of x-and y-values can be displayed. We will use a table to evaluate the formula for the given values.

Once the formula is entered, press $\boxed{\text{2nd}}$ $\boxed{\text{TBLSET}}$ to display the table set-up screen. (TBLSET is the second function associated with the $\boxed{\text{WINDOW}}$ key.) You can choose to supply the x-values yourself or you can set the calculator to supply them. To choose the x-values yourself, set "Indpnt" to "Ask" by using the $\boxed{\triangledown}$ and $\boxed{\triangleright}$ keys to position the cursor over "Ask" and then pressing $\boxed{\text{ENTER}}$. "Depend" should be set to "Auto." In Ask mode the graphing calculator disregards the settings of TblStart and ΔTbl.

Now press [2nd] [TABLE] to view the table. (TABLE is the second operation associated with the [GRAPH] key.) Values for x can be entered in the X-column of the table and the corresponding values of y_1 will be displayed in the Y_1-column. To enter 800, 1500, 2400, and 3600, press 8 0 0 [ENTER] 1 5 0 0 [ENTER] 2 4 0 0 [ENTER] 3 6 0 0 [ENTER]. The down arrow key [▽] can be pressed instead of [ENTER] if desired.

We see that $y_1 = 24,000$ when $x = 800$, $y_1 = 45,000$ when $x = 1500$, $y_1 = 72,000$ when $x = 2400$, and $y_1 = 108,000$ when $x = 3600$, so the furnace outputs for 800 ft^2, 1500 ft^2, 2400 ft^2, and 3600 ft^2 are 24,000 Btu's, 45,000 Btu's, 72,000 Btu's, and 108,000 Btu's, respectively.

THE TABLE FEATURE: AUTO MODE

Section 2.4, Example 8 Village Stationers wants the customer service team to be able to look up the cost c of merchandise being returned when only the total amount paid T (including tax) is shown on the receipts. Use the formula $c = T/1.05$, developed in Example 7, to create a table of values showing cost given a total amount paid. Assume that the least possible sale is \$0.21.

We will use a graphing calculator to create the table of values. To enter the formula, we first replace c with y and T with x. Then we enter the formula on the equation-editor screen as $y = x/1.05$. Be sure that the Plots are turned off and that any previous entries are cleared. (See page 9 of this manual for the procedure for turning off the Plots and clearing equations.)

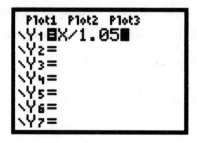

TI-83/83 Plus

We will create a table in which the calculator supplies the x-values beginning with a value we specify and continuing by adding a value we specify to the preceding value for x. We will begin with an x-value of 0.21 (corresponding to $0.21) and choose successive increases of 0.01 (corresponding to $0.01). To do this, first press $\boxed{\text{2nd}}$ $\boxed{\text{TBLSET}}$ to access the TABLE SETUP window. If "Indpnt" is set to "Auto," the calculator will supply values for x, beginning with the value specified as TblStart and continuing by adding the value of ΔTbl to the preceding value for x. Press $\boxed{.}$ $\boxed{2}$ $\boxed{1}$ $\boxed{\triangledown}$ $\boxed{.}$ $\boxed{0}$ $\boxed{1}$ to select a beginning x-value of 0.21 and an increment of 0.01. The "Indpnt" and "Depend" settings should both be "Auto." If either is not, use the $\boxed{\triangledown}$ key to position the blinking cursor over "Auto" on that line and then press $\boxed{\text{ENTER}}$. To display the table press $\boxed{\text{2nd}}$ $\boxed{\text{TABLE}}$.

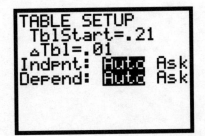

We can change the number of decimal places that the graphing calculator will display to 2 so that the costs are rounded to the nearest cent. To do this, press $\boxed{\text{MODE}}$ to access the MODE screen. Then press $\boxed{\triangledown}$ to move the cursor to the second line and press the $\boxed{\triangleright}$ key three times to highlight 2. (This assumes that "Float" was previously selected.) Finally, press $\boxed{\text{ENTER}}$ to select two decimal places. To return to the table, press $\boxed{\text{2nd}}$ $\boxed{\text{TABLE}}$.

Use the $\boxed{\triangledown}$ and $\boxed{\triangle}$ keys to scroll through the table.

Before proceeding, return to the MODE screen and reselect "Float." This allows the number of decimal places to vary, or float, according to the computation being performed.

In **Section 2.5, Example 2**, two equations are entered on the equation-editor screen. The first equation, $y_1 = x + 1$, can be entered as described on page 9 of this manual. Then, to enter $y_2 = x + (x+1)$ press either $\boxed{\triangledown}$ or $\boxed{\text{ENTER}}$ to position the cursor beside "$Y_2 =$." Now enter the right-hand side of the equation.

```
Plot1  Plot2  Plot3
\Y1■X+1
\Y2■X+(X+1)
\Y3=
\Y4=
\Y5=
\Y6=
\Y7=
```

Chapter 3
Introduction to Graphing and Functions

If you selected 2 for the number of decimal places in Section 2.4, Example 8, and have not yet returned to the MODE screen to reselect "Float," do so now. In Float mode the number of decimal places varies, or floats, according to the computation being performed.

SETTING THE VIEWING WINDOW

The viewing window is the portion of the coordinate plane that appears on the graphing calculator's screen. It is defined by the minimum and maximum values of x and y: Xmin, Xmax, Ymin, and Ymax. The notation [Xmin, Xmax, Ymin, Ymax] is used in the text to represent these window settings or dimensions. For example, [−12, 12, −8, 8] denotes a window that displays the portion of the x-axis from −12 to 12 and the portion of the y-axis from −8 to 8. In addition, the distance between tick marks on the axes is defined by the settings Xscl and Yscl. In this manual Xscl and Yscl will be assumed to be 1 unless noted otherwise. The setting Xres sets the pixel resolution. We usually select Xres = 1. The window corresponding to the settings [−20, 30, −12, 20], Xscl = 5, Yscl = 2, Xres = 1, is shown below. Note that all of the entries on the equation-editor screen have been cleared and the plots have been turned off in order to show this window. (See page 9 of this manual for the procedure for clearing entries and turning off the plots.)

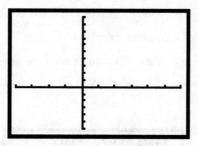

Press the WINDOW key on the top row of the keypad to display the current window settings on your calculator. The standard settings are shown below.

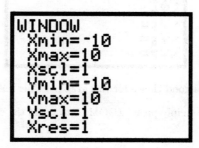

Section 3.1, Example 5 Set up a $[-100, 100, -5, 5]$ viewing window on a graphing calculator, choosing appropriate scales for the axes.

To change a setting, position the cursor beside the setting you wish to change and enter the new value. We will enter the given settings and let Xscl = 01 and Yscl =1. The choice of scales for the axes may vary. We select scaling that will allow some space between tick marks. If a scale that is too small is chosen, the tick marks will blend and blur. To change the settings to $[-100, 100, -5, 5]$, Xscl = 10, press $\boxed{\text{WINDOW}}$ $\boxed{(-)}$ 1 0 0 $\boxed{\text{ENTER}}$ 1 0 0 $\boxed{\text{ENTER}}$ 1 0 $\boxed{\text{ENTER}}$ $\boxed{(-)}$ 5 $\boxed{\text{ENTER}}$ 5 $\boxed{\text{ENTER}}$ 1 $\boxed{\text{ENTER}}$. The $\boxed{\triangledown}$ key may be used instead of $\boxed{\text{ENTER}}$ after typing each window setting. To see the window with these settings, press the $\boxed{\text{GRAPH}}$ key on the top row of the keypad.

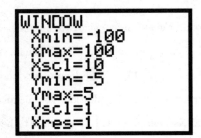
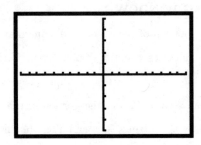

QUICK TIP: To return quickly to the standard window setting $[-10, 10, -10, 10]$, Xscl = 1, Yscl = 1, press $\boxed{\text{ZOOM}}$ 6.

GRAPHING EQUATIONS

After entering an equation and setting a viewing window, you can view the graph of an equation.

Section 3.2, Example 4 Graph $y = 2x$ using a graphing calculator.

Press $\boxed{Y =}$ to access the equation-editor screen. Turn off the plots and clear any previous entries. (See page 9 of this manual.) Then use $\boxed{\triangle}$ or $\boxed{\triangledown}$ to move the cursor to the top line beside "$Y_1 =$." Now press 2 $\boxed{X, T, \Theta, n}$ to enter the right-hand side of the equation on the "Y =" screen.

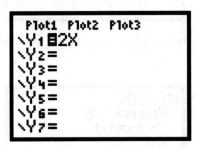

The standard $[-10, 10, -10, 10]$ window is a good choice for this graph. Either enter these dimensions in the WINDOW screen and then press $\boxed{\text{GRAPH}}$ to see the graph or simply press $\boxed{\text{ZOOM}}$ 6 to select the standard window and see the graph.

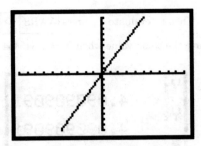

SOLVING EQUATIONS GRAPHICALLY; THE INTERSECT METHOD

We can use the Intersect feature from the CALC menu to solve equations.

Section 3.3, Example 2 Solve using a graphing calculator: $-\frac{3}{4}x + 6 = 2x - 1$.

On the equation editor screen, clear any existing entries and then enter $y_1 = -(3/4)x + 6$ and $y_2 = 2x - 1$. Although the parentheses in y_1 are not necessary, they make the equation easier to read on the equation-editor screen. Press ZOOM 6 to graph these equations in the standard viewing window. The solution of the equation $-\frac{3}{4}x + 6 = 2x - 1$ is the first coordinate of the point of intersection of the graphs. To use the Intersect feature to find this point, first press 2nd CALC 5 to select Intersect from the CALC menu. (CALC is the second operation associated with the TRACE key on the top row of the keypad.) The query "First curve?" appears at the bottom of the screen. The blinking cursor is positioned on the graph of y_1. This is indicated by the notation "$Y_1 = -\frac{3}{4}x + 6$" in the upper left-hand corner of the screen. Press ENTER to indicate that this is the first curve involved in the intersection. Next the query "Second curve?" appears at the bottom of the screen. The blinking cursor is now positioned on the graph of y_2 and the notation "$Y_2 = 2x - 1$" should appear in the top left-hand corner of the screen. Press ENTER to indicate that this is the second curve. We identify the curves for the calculator since we could have as many as ten graphs on the screen at once. After we identify the second curve, the query "Guess?" appears at the bottom of the screen. Use the right and left arrow keys to move the blinking cursor close to the point of intersection of the graphs. This provides the calculator with a guess as to the coordinates of this point. We do this since some pairs of curves can have more than one point of intersection. When the cursor is positioned, press ENTER a third time. Now the coordinates of the point of intersection appear at the bottom of the screen. We see that $x = 2.5454545$, so the solution of the equation is 2.5454545.

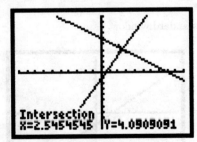

We can check the solution by evaluating both sides of the equation $-\frac{3}{4}x + 6 = 2x - 1$ for this value of x. The first coordinate of the point of intersection has automatically been stored as x in the calculator, so we evaluate y_1 and y_2 for this value of x. First press 2nd QUIT to go to the home screen. Then to evaluate Y_1 press VARS ▷ 1 1 ENTER. To evaluate Y_2 press VARS ▷ 1 2 ENTER. We see that Y_1 and Y_2 have the same value when X = 2.5454545, so the solution checks.

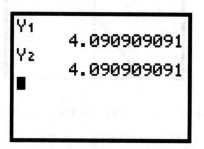

Note that although the procedure above verifies that 2.5454545 is the solution, this number is actually an approximation of the solution. In some cases the calculator will give an exact solution. Since the x-coordinate of the point of intersection is stored in the calculator as X, we can find an exact solution by converting X to fraction notation. This can be done from the home screen by pressing X,T,Θ,n MATH 1 ENTER or X,T,Θ,n MATH ENTER ENTER. We see that the exact solution is $\frac{28}{11}$.

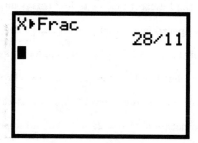

SOLVING EQUATIONS GRAPHICALLY; THE ZERO METHOD

As an alternative to the Intersect method, we can use the Zero feature from the CALC menu to solve equations.

Section 3.3, Example 7 Solve $3 - 8x = 5 - 7x$ using the zero method.

First we get zero on one side of the equation:
$$3 - 8x = 5 - 7x$$
$$-2 - 8x = -7x$$
$$-2 - x = 0.$$
Then we graph $y = -2 - x$. We will use the standard window.

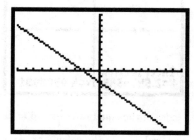

The first coordinate of the x-intercept of the graph is the zero of $y = -2 - x$ and thus is the solution of the original equation.

TI-83/83 Plus

Press [2nd] [CALC] 2 to select Zero from the CALC menu. We are prompted to select a left bound. This means that we must choose an x-value that is to the left of the first coordinate of the x-intercept. This can be done by using the left and right arrow keys to move the cursor to a point on the graph that is to the left of the intercept or by keying in an x-value that is less than the first coordinate of the intercept.

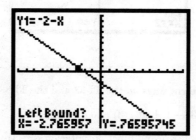

Once this is done, press [ENTER]. Now we are prompted to select a right bound that is to the right of the x-intercept. Again we can use the arrow keys or key in a value.

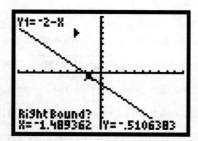

Press [ENTER] again. Finally, we are prompted to make a guess as to the value of the zero. Move the cursor to a point close to the zero or key in a value.

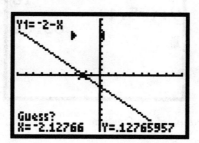

Press [ENTER] a third time. We see that $y = 0$ when $x = -2$, so -2 is the zero of $y = -2 - x$ and is thus the solution of the original equation, $3 - 8x = 5 - 7x$.

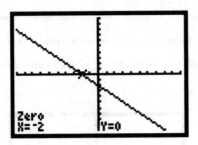

EVALUATING A FUNCTION

Function values can be found in several different ways on the TI-83 and the TI-83 Plus.

Section 3.4, Example 5 For $f(a) = 2a^2 - 3a + 1$, find $f(3)$ and $f(-5.1)$.

One method for finding function values involves using function notation directly. To do this, first press $\boxed{Y=}$ and enter the function on the equation-editor screen. Mentally replace a with x and $f(a)$ with Y_1. Then enter $Y_1 = 2x^2 - 3x + 1$. Now, to find $f(3)$, or $Y_1(3)$, directly first press $\boxed{\text{2nd}}$ $\boxed{\text{QUIT}}$ to go to the home screen. Then press $\boxed{\text{VARS}}$ $\boxed{\triangleright}$ 1 1 $\boxed{(}$ 3 $\boxed{)}$ $\boxed{\text{ENTER}}$. We see that $Y_1(3) = 10$, or $f(3) = 10$.

To find $f(-5.1)$, or $Y_1(-5.1)$, we can repeat the previous procedure using -5.1 in place of 3, or we can edit the previous entry. To edit, copy the entry $Y_1(3)$ to the home screen by pressing $\boxed{\text{2nd}}$ $\boxed{\text{ENTRY}}$. (ENTRY is the second operation associated with the $\boxed{\text{ENTER}}$ key.) Now replace 3 with -5.1 by first pressing $\boxed{\triangleleft}$ $\boxed{\triangleleft}$ to position the cursor over the 3. Then press $\boxed{(-)}$ to overwrite the 3 with the negative symbol. To insert 5.1 after this symbol press $\boxed{\text{2nd}}$ $\boxed{\text{INS}}$ 5 $\boxed{.}$ 1. (INS, for "insert," is the second operation associated with the $\boxed{\text{DEL}}$ key.) Finally press $\boxed{\text{ENTER}}$ to find that $Y_1(-5.1) = 68.32$, or $f(-5.1) = 68.32$.

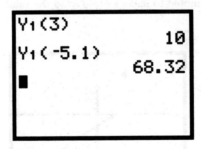

We can also find function values from the graph of the function.

Section 3.4, Example 6 Find $g(2)$ for $g(x) = 2x - 5$.

First press $\boxed{Y=}$ to go to the equation editor screen and then clear any entries that are present. Also be sure that the Stat Plots are turned off. (See page 9 of this manual for instructions for clearing equations and turning off Stat Plots.) Now enter $Y_1 = 2x - 5$ and press $\boxed{\text{ZOOM}}$ 6 to graph this function in the standard viewing window. We will use the Value feature from the CALC menu to find the value of Y_1 when $x = 2$. This is $g(2)$. Press $\boxed{\text{2nd}}$ $\boxed{\text{CALC}}$ 1 to select Value. (CALC is the second operation associated with the $\boxed{\text{TRACE}}$ key in the top row of the keypad.) Now you must supply the value of x as indicated by the blinking cursor at the bottom of the screen beside X =. Press 2 $\boxed{\text{ENTER}}$. We now see X = 2, Y = -1 at the bottom of the screen, so $g(2) = -1$.

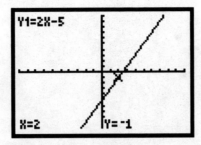

When using the Value feature, note that the x-value entered must be in the viewing window. That is, x must be a number between Xmin and Xmax.

Chapter 4
Linear Equations, Inequalities, and Graphs

SQUARING THE VIEWING WINDOW

Section 4.4, Example 9 Determine whether the lines given by the equations $3x - y = 7$ and $x + 3y = 1$ are perpendicular, and check by graphing.

In the text each equation is solved for y in order to determine the slopes of the lines. We have $y = 3x - 7$ and $y = -\frac{1}{3}x + \frac{1}{3}$. Since $3\left(-\frac{1}{3}\right) = -1$, we know that the lines are perpendicular. To check this, we graph $Y_1 = 3x - 7$ and $Y_2 = -\frac{1}{3}x + \frac{1}{3}$. The graphs are shown on the right below in the standard viewing window.

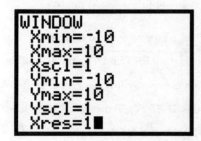
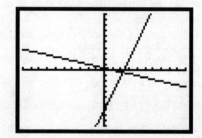

Note that the graphs do not appear to be perpendicular. This is due to the fact that, in the standard window, the distance between tick marks on the y-axis is about 2/3 the distance between tick marks on the x-axis. It is often desirable to choose window dimensions for which these distances are the same, creating a "square" window. On the TI-83 and TI-83 Plus, any window in which the ratio of the length of the y-axis to the length of the x-axis is 2/3 will produce this effect.

This can be accomplished by selecting dimensions for which Ymax − Ymin = $\frac{2}{3}$(Xmax − Xmin). For example, the windows $[-12, 12, -8, 8]$ and $[-6, 6, -4, 4]$ are square. When we change the window dimensions to $[-12, 12, -8, 8]$ and press GRAPH, the graphs now appear to be perpendicular as shown on the right below. Instead of entering window dimensions, we could also press ZOOM 5 and the calculator will select a square window.

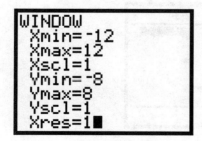
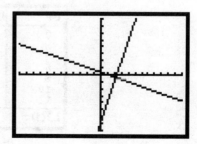

ENTERING AND PLOTTING DATA; LINEAR REGRESSION

We can use the Linear Regression feature in the STAT CALC menu to fit a linear equation to a set of data.

Section 4.5, Example 9 The amount of paper recovered in the United States for various years is shown in the following table.

Years	Amount of Paper Recovered (in millions of tons)
1988	26.2
1990	29.1
1992	34.0
1994	39.7
1996	43.1
1998	45.1
2000	49.4

(a) Fit a linear function to the data.

(b) Graph the function and use it to estimate the amount of paper that will be recovered in 2003.

(a) We will enter the coordinates of the ordered pairs on the STAT list editor screen. To clear any existing lists press $\boxed{\text{STAT}}$ 4 $\boxed{\text{2nd}}$ $\boxed{L_1}$, $\boxed{\text{2nd}}$ $\boxed{L_2}$, $\boxed{\text{2nd}}$ $\boxed{L_3}$, $\boxed{\text{2nd}}$ $\boxed{L_4}$, $\boxed{\text{2nd}}$ $\boxed{L_5}$, $\boxed{\text{2nd}}$ $\boxed{L_6}$ $\boxed{\text{ENTER}}$. (L_1 through L_6 are the second operations associated with the numeric keys 1 through 6.) The lists can also be cleared by first accessing the STAT list editor screen by pressing $\boxed{\text{STAT}}$ $\boxed{\text{ENTER}}$ or $\boxed{\text{STAT}}$ 1. These keystrokes display the STAT EDIT menu and then select the Edit option from that menu. Then, for each list that contains entries, use the arrow keys to move the cursor to highlight the name of the list at the top of the column and press $\boxed{\text{CLEAR}}$ $\boxed{\triangledown}$ or $\boxed{\text{CLEAR}}$ $\boxed{\text{ENTER}}$.

Once the lists are cleared, we can enter the coordinates of the points. We will enter the first coordinates (x-coordinates), as the number of years since 1988, in L_1 and the second coordinates (y-coordinates), in millions of tons, in L_2. With the STAT list editor screen displayed as described above, position the cursor at the top of column L_1, below the L_1 heading. To enter 0 (for 1988) press 0 $\boxed{\text{ENTER}}$. Continue entering the x-values 2, 4, 6, 8, 10, and 12, each followed by $\boxed{\text{ENTER}}$. The entries can be followed by $\boxed{\triangledown}$ rather than $\boxed{\text{ENTER}}$ if desired. Press $\boxed{\triangleright}$ to move to the top of column L_2. Enter the y-values 26.2, 29.1, 34.0, 39.7, 43.1, 45.1, and 49.4 in succession, each followed by $\boxed{\text{ENTER}}$ or $\boxed{\triangledown}$. Note that the coordinates of each point must be in the same position in both lists.

```
L1      L2      L3    3
0       26.2    ------
2       29.1
4       34
6       39.7
8       43.1
10      45.1
12      49.4
L3(1)=
```

press $\boxed{Y=}$ to go to the equation-editor screen and clear any equations that are currently entered. (See page 9 of this
If you wish, instead of clearing an equation, you can deselect it. To do this, position the cursor over the = sign and

press ENTER. Note that the = sign is no longer highlighted, indicating that the equation has been deselected. The graph of an equation that has been deselected will not appear when GRAPH is pressed. A deselected equation can be selected again by positioning the cursor over the = sign and pressing ENTER. Note that the = sign is once again highlighted.

Now use the graphing calculator's linear regression feature to fit a linear equation to the data. Press STAT ▷ 4 ENTER to select LinReg($ax + b$) from the STAT CALC menu and to display the coefficients a and b of the regression equation $y = ax + b$. We see that the regression equation is $y = 1.976785714x + 26.225$.

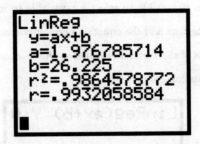

If the diagnostics have been turned on in your calculator, values for r^2 and r will also be displayed. These numbers indicate how well the regression line fits the data. For the remainder of this manual, regression will be done with the diagnostics turned off.

If you wish to select DiagnosticOn mode, press 2nd CATALOG and use ▽ to position the triangular selection cursor beside DiagnosticOn. To alleviate the tedium of scrolling through many items to reach DiagnosticOn, press D after pressing 2nd CATALOG to move quickly to the first catalog item that begins with the letter D. (D is the ALPHA operation associated with the x^{-1} key.) Then use ▽ to scroll to DiagnosticOn. Note that it is not necessary to press ALPHA before D when the catalog is displayed. Press ENTER to paste this instruction to the home screen and then press ENTER a second time to set the mode. To select DiagnosticOff mode, press 2nd CATALOG, position the selection cursor beside DiagnosticOff, press ENTER to paste this instruction to the home screen, and then press ENTER again to set this mode.

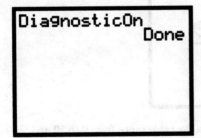
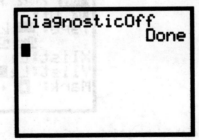

Immediately after the regression equation is found it can be copied to the equation-editor screen as Y_1. Note that any previous entry in Y_1 must have been cleared rather than deselected. Press Y = and position the cursor beside Y_1. Then press VARS 5 ▷ ▷ 1. These keystrokes select Statistics from the VARS menu, then select the EQ (Equation) submenu, and finally select the RegEq (Regression Equation) from this submenu.

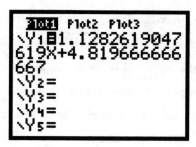

Before the regression equation is found, it is possible to select a *y*-variable to which it will be stored on the equation editor screen. After the data have been stored in the lists and the equation previously entered as Y_1 has been cleared, press STAT ▷ 4 VARS ▷ 1 1 ENTER. The coefficients of the regression equation will be displayed on the home screen, and the regression equation will also be stored as Y_1 on the equation-editor screen.

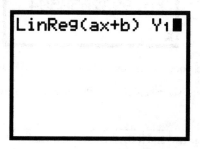

(b) Now we will graph the regression equation. In order to see the data points along with the graph of the equation we will turn on and define a Stat Plot. To do this, first press 2nd STATPLOT to go to the STAT PLOTS screen. Press ENTER to select Plot 1 and then position the cursor over On and press ENTER to turn on Plot 1. Next select the scatter diagram for Type, L_1 for Xlist, L_2 for Ylist, and the box for the Mark as shown below. To select Type and Mark, position the cursor over the desired selection and press ENTER. Use the L_1 and L_2 keys (associated with the 1 and 2 numeric keys) to select Xlist and Ylist.

To select the dimensions of the viewing window notice that the years in the table range from 0 to 12 and the number of millions of tons of paper ranges from 26.2 to 49.4. We want to select dimensions that will include all of these values. One good choice is [0, 15, 0, 60], Yscl = 10. Enter these dimensions in the WINDOW screen.

Once the equation has been entered on the equation-editor screen as described above, press GRAPH to graph the regression line on the same axes as the data.

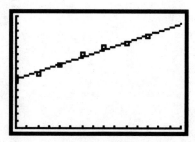

To estimate the amount of paper that will be recovered in 2003, evaluate the regression equation for $x = 15$. (2003 is 15 years after 1988.) Use any of the methods for evaluating a function presented earlier in this chapter. (See page 18.) We will use the Value feature from the CALC menu.

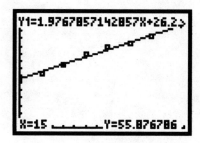

When $x = 15, y \approx 55.9$, so we estimate that about 55.9 million tons of paper will be recovered in 2003.

THE TRACE FEATURE

Section 4.6, Example 10 Find the domain of $f + g$ if $f(x) = \sqrt{2x - 5}$ and $g(x) = \sqrt{x + 1}$.

In the text it is found that the domain of $f + g$ is $\left\{x \middle| x \geq \dfrac{5}{2}\right\}$, or $\left[\dfrac{5}{2}, \infty\right)$. This can be confirmed, at least approximately, by tracing the graph of $f + g$. First graph $y = \sqrt{2x - 5} + \sqrt{x + 1}$. We will use the standard window. Now press $\boxed{\text{TRACE}}$ and use the right and left arrow keys to move the cursor along the curve. We see that no y-values are given for x-values less than 2.5, or $\dfrac{5}{2}$. This indicates that x-values less than 2.5 are not in the domain of $f + g$. As we move to the right, ordered pairs appear to extend without bound confirming that the domain of $f + g$ is $\left\{x \middle| x \geq \dfrac{5}{2}\right\}$, or $\left[\dfrac{5}{2}, \infty\right)$.

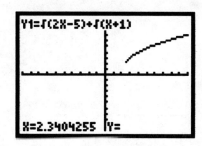 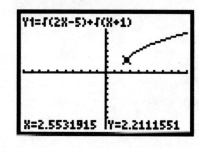

Chapter 5
Polynomials

CHECKING OPERATIONS ON POLYNOMIALS

A graphing calculator can be used to check operations on polynomials.

Section 5.3, Interactive Discovery Check the addition $(-3x^3 + 2x - 4) + (4x^3 + 3x^2 + 2) = x^3 + 3x^2 + 2x - 2$.

There are several ways in which we can use a graphing calculator to check this result. One of these is to compare the graphs of $Y_1 = (-3x^3 + 2x - 4) + (4x^3 + 3x^2 + 2)$ and $Y_2 = x^3 + 3x^2 + 2x - 2$. This is most easily done when different graph styles are used for the graphs.

Seven graph styles can be selected on the equation-editor screen of the TI-83 and the TI-83 Plus. The **path graph style** can be used, along with the line style, to determine whether graphs coincide. To use graphs to check the addition in Example 9, first press $\boxed{\text{MODE}}$ to determine whether Sequential mode is selected. If it is not, position the blinking cursor over Sequential and then press $\boxed{\text{ENTER}}$. Next, on the Y = screen, enter $Y_1 = (-3x^3 + 2x - 4) + (4x^3 + 3x^2 + 2)$ and $Y_2 = x^3 + 3x^2 + 2x - 2$. We will select the line graph style for Y_1 and the path style for Y_2. To select these graph styles use $\boxed{\triangleleft}$ to position the cursor over the icon to the left of the equation and press $\boxed{\text{ENTER}}$ repeatedly until the desired style icon appears as shown on the right below.

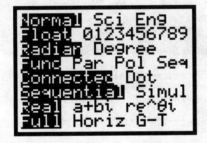 

The calculator will graph Y_1 first as a solid line. Then Y_2 will be graphed as the circular cursor traces the leading edge of the graph, allowing us to determine visually whether the graphs coincide. In this case, the graphs appear to coincide, so the factorization is probably correct.

We can also check the addition by **subtracting** the result from the original sum. With Y_1 and Y_2 entered as described above, position the cursor beside "$Y_3 =$" and use the Y-VARS menu to enter $Y_3 = Y_1 - Y_2$ by pressing $\boxed{\text{VARS}}$ $\boxed{\triangleright}$ 1 1 $\boxed{(-)}$ $\boxed{\text{VARS}}$ $\boxed{\triangleright}$ 1 2. If the expressions for Y_1 and Y_2 are equivalent, the graph of Y_3 will be $y = 0$, or the x-axis. Since we are interested only in the values of Y_3, deselect Y_1 and Y_2 as described on page 22 of this manual and select the path graph style for Y_3 as described above.

Now press GRAPH and determine if the graph of Y_3 is traced over the x-axis. Since it is, the sum is correct.

We can use a table of values to **compare values** of Y_1 and Y_2. If the expressions for Y_1 and Y_2 are the same for each given x-value, the result checks. If you deselected Y_1 and Y_2 to check the sum using subtraction as described above, select them again now. Then look at a table set in Auto mode. Since the values of Y_1 and Y_2 are the same for each given x-value, the result checks. Scrolling through the table to look at additional values makes this conclusion more certain.

We can also check the sum using a **horizontal split-screen**. The top half of the screen displays the graph and, in this case, we will use the bottom half to display a table of values.

First enter Y_1 and Y_2 as described above. Then, to select the horizontal split-screen option, press MODE to access the Mode screen. Position the cursor over Horiz on the last line and press ENTER. Press GRAPH to see the graph in the top half of the split screen, and press 2nd TABLE to see two lines of the table of values for y_1 and y_2 below the graph. We show these functions graphed in the standard window. We show a table with TblStart $= -3$, ΔTbl $= 1$, and Indpnt and Depend both set on Auto. The graphs appear to coincide. In addition, as we scroll through the table, we see that the values of Y_1 and Y_2 are the same for any given x-value, so the sum is probably correct.

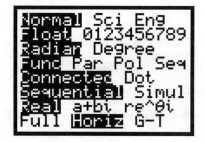 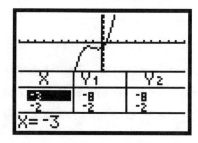

The lower half of the split screen can also display four lines of the home screen, four lines of the equation-editor screen, two rows of the Stat list editor screen, or three settings of the Window screen instead of two lines of the table. To change from the table to the home screen, press 2nd QUIT. Display the equation-editor screen by pressing Y=, the Stat list editor by pressing STAT 1, or the Window screen by pressing WINDOW.

In order to return to a full-screen graph, table, equation-editor, Stat list editor, or window screen, return to the Mode screen and select Full.

We can also use a **vertical split screen** to check the addition. First enter Y_1, Y_2, and Y_3 as described above and deselect Y_1 and Y_2. Then press MODE to display the Mode screen, position the cursor over G-T on the last line, and press ENTER. Now press GRAPH to see the graph of Y_3 on the left side of the screen and a table of values for Y_3 on the right side. As we did above, we use a table set in Auto mode with TblStart $= -3$ and ΔTbl $= 1$. Since the graph appears to be $y = 0$, or the x-axis, and all of the Y_3-values in the table are 0, we confirm that the result is correct.

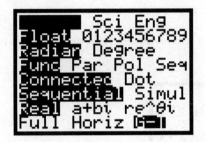
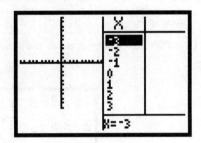

In order to return to full-screen mode, return to the Mode screen and select Full.

EVALUATING POLYNOMIALS IN SEVERAL VARIABLES

Section 5.6, Example 1 Evaluate the polynomial $4 + 3x + xy^2 + 8x^3y^3$ for $x = -2$ and $y = 5$.

To evaluate a polynomial in two or more variables, substitute numbers for the variables. This can be done by substituting values for x and y directly or by storing the values of the variables in the calculator. The procedure for direct substitution is described on page 3 of this manual.

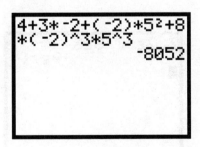

To evaluate the polynomial function by first storing -2 as x and 5 as y, we proceed as follows. To store -2 as x, press (−) 2 STO▷ X, T, Θ, n ENTER. Now store 5 as y. Press 5 STO▷ ALPHA Y ENTER. ALPHA is the green key in the left-hand column of the keypad, and Y is the alpha, or letter, operation associated with the 1 numeric key. Next enter the algebraic expression. Press 4 + 3 X, T, Θ, n + X, T, Θ, n ALPHA Y ∧ 2 + 8 X, T, Θ, n ∧ 3 ALPHA Y ∧ 3. Finally, press ENTER to find the value of the expression.

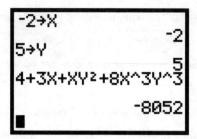

SCIENTIFIC NOTATION

To enter a number in scientific notation, first type the decimal portion of the number; then press $\boxed{\text{2nd}}$ $\boxed{\text{EE}}$ (EE is the second operation associated with the $\boxed{,}$ key.); finally type the exponent, which can be at most two digits. For example, to enter 1.789×10^{-11} in scientific notation, press 1 $\boxed{.}$ 7 8 9 $\boxed{\text{2nd}}$ $\boxed{\text{EE}}$ $\boxed{(-)}$ 1 1 $\boxed{\text{ENTER}}$. To enter 6.084×10^{23} in scientific notation, press 6 $\boxed{.}$ 0 8 4 $\boxed{\text{2nd}}$ $\boxed{\text{EE}}$ 2 3 $\boxed{\text{ENTER}}$. The decimal portion of each number appears before a small E while the exponent follows the E.

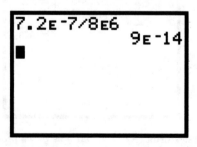

The calculator can be used to perform computations in scientific notation.

Section 5.8, Example 9 Use a graphing calculator to check the computation $(7.2 \times 10^{-7}) \div (8.0 \times 10^6) = 9.0 \times 10^{-14}$.

We enter the computation in scientific notation. Press 7 $\boxed{.}$ 2 $\boxed{\text{2nd}}$ $\boxed{\text{EE}}$ $\boxed{(-)}$ 7 $\boxed{\div}$ 8 $\boxed{\text{2nd}}$ $\boxed{\text{EE}}$ 6 $\boxed{\text{ENTER}}$. We have 9×10^{-14}, which checks.

Chapter 6
Polynomials and Factoring

THE INTERSECT AND ZERO METHODS

Section 6.1, Example 1 Solve: $x^2 = 6x$.

To solve this equation by finding the x-coordinates of the points of intersection of $y_1 = x^2$ and $y_2 = 6x$, see the intersect method described on page 15 of this manual.

To solve this equation by writing it as $x^2 - 6x = 0$, graphing $f(x) = x^2 - 6x$, and then finding the values of x for which $f(x) = 0$, or the first coordinates of the x-intercepts of the graph, see the zero method described on page 16 of this manual.

POLYNOMIAL REGRESSION

The TI-83 and TI-83 Plus have the capability to use regression to fit nonlinear polynomial equations to data.

Section 6.7, Example 4(a) The number of bachelor's degrees earned in the biological and life sciences for various years is shown in the following table. Fit a polynomial function of degree 3 (cubic) to the data.

Year	Number of Bachelor's Degrees Earned in Biological/Life Science
1971	35,743
1976	54,275
1980	46,370
1986	38,524
1990	37,204
1994	51,383
2000	63,532

First enter the data with the number of years since 1970 in L_1 and the number of bachelor's degrees earned, in thousands, in L_2. (See page 22 of this manual for the procedure to follow.) We select cubic regression, denoted CubicReg, from the STAT CALC menu. Press $\boxed{\text{STAT}}$ $\boxed{\triangleright}$ 6 $\boxed{\text{ENTER}}$. (Note that if the data are entered in a combination of lists other than L_1 and L_2 these lists must be specified. See page 36 of this manual.) The calculator returns the coefficients for a cubic function of the form $f(x) = ax^3 + bx^2 + cx + d$. Rounding the coefficients to the nearest thousandth, we have $f(x) = 0.009x^3 - 0.371x^2 + 4.176x + 34.415$.

We can use methods discussed earlier in this manual to estimate and predict function values and to find the year or years in which a specific function value occurs.

Chapter 7
Rational Expressions, Equations, and Functions

GRAPHING IN DOT MODE

Consider the graph of the function $T(t) = \dfrac{t^2 + 5t}{2t + 5}$ in Section 7.1, Example 1. Enter $y = (x^2 + 5x)/(2x + 5)$ and graph it in the window $[-5, 5, -5, 5]$.

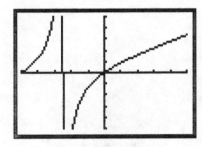

Note that a vertical line that is not part of the graph appears on the screen along with the two branches of the graph. The reason for this is discussed in the text.

This line will not appear if we change from Connected mode to Dot mode. Access the Mode screen by pressing $\boxed{\text{MODE}}$. Then move the cursor to Dot on the fifth line and press $\boxed{\text{ENTER}}$. Now press $\boxed{\text{GRAPH}}$ to see the graph of the function in Dot mode.

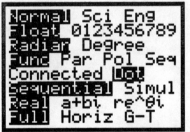 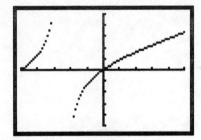

We can also select Dot mode by selecting the "dot" GraphStyle on the equation-editor screen. If the function $T(t) = \dfrac{t^2 + 5t}{2t + 5}$ is entered as $y_1 = (x^2 + 5x)/(2x + 5)$, for instance, position the cursor over the GraphStyle icon to the left of Y_1 and press $\boxed{\text{ENTER}}$ repeatedly until the "dot" icon appears. If the "line" icon was previously selected, $\boxed{\text{ENTER}}$ must be pressed six times to select the "dot" style.

It should be noted that, when an equation is cleared on the equation-editor screen of the TI-83 or the TI-83 Plus, the GraphStyle returns to "line" (or connected mode) regardless of the mode selected on the MODE screen.

Chapter 8
Systems of Linear Equations and Problem Solving

SOLVING SYSTEMS OF EQUATIONS GRAPHICALLY

We can use the Intersect feature from the CALC menu on the TI-83 and the TI-83 Plus to solve a system of two equations in two variables.

Section 8.1, Example 4(a) Solve graphically:

$$y - x = 1,$$
$$y + x = 3.$$

We graph the equations in the same viewing window and then find the coordinates of the point of intersection. Remember that equations must be entered in "$y =$" form on the equation-editor screen, so we solve both equations for y. We have $y = x + 1$ and $y = -x + 3$. Enter these equations, graph them in the standard viewing window, and find their point of intersection as described on page 15 of this manual. We see that the solution of the system of equations is $(1, 2)$.

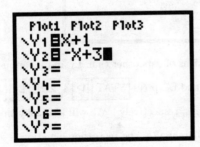 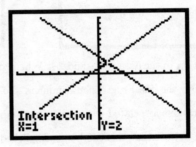

MODELS

Sometimes we model two situations with linear equations and then want to find the point of intersection of their graphs.

Section 8.1, Example 6 (d), (e) The numbers of U. S. travelers to Canada and to Europe are listed in the following table.

Year	U. S. Travelers to Canada (in millions)	U. S. Travelers to Europe (in millions)
1992	11.8	7.1
1994	12.5	8.2
1996	12.9	8.7
1998	14.9	11.1
2000	15.1	13.4

(d) Use linear regression to find two linear equations that can be used to estimate the number of U. S. travelers to Canada and Europe, in millions, x years after 1990.

(e) Use the equations found in part (d) to estimate the year in which the number of U. S. travelers to Europe will be the same as the number of U. S. travelers to Canada.

(d) Enter the data in STAT lists as described on page 22 of this manual. We will express the years as the number of years after 1990 (in other words, 1990 is year 0) and enter them in L_1. Then enter the number of travelers to Canada, in millions, in L_2 and the number of travelers to Europe, in millions, in L_3.

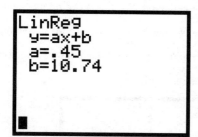

Now use linear regression to fit a linear function to the data in L_1 and L_2. The function should also be copied to the equation editor screen. We will copy it as Y_1. See pages 23 and 24 of this manual for the procedure to follow. We get $y_1 = 0.45x + 10.74$.

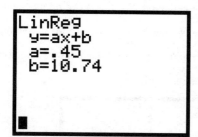 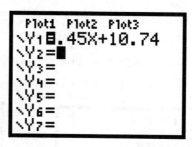

Next we fit a linear function to the data in L_1 and L_3. When combinations of lists other than L_1 and L_2 are used, the names of the lists must be entered after the linear regression command. To use L_1 and L_3, press $\boxed{\text{STAT}}$ $\boxed{\triangleright}$ $\boxed{4}$ $\boxed{\text{2nd}}$ $\boxed{L_1}$ $\boxed{,}$ $\boxed{\text{2nd}}$ $\boxed{L_3}$. (L_1 and L_3 are the second operations associated with the 1 and 3 numeric keys, respectively.) We will also copy this function to the equation-editor screen as Y_2. Note that this can be accomplished immediately after the regression equation is found by pressing $\boxed{Y=}$, positioning the cursor beside "$Y_2 =$", clearing the existing entry if one exists, and then pressing $\boxed{\text{VARS}}$ $\boxed{5}$ $\boxed{\triangleright}$ $\boxed{\triangleright}$ $\boxed{1}$. It can also be done before the regression equation is found by pressing $\boxed{,}$ $\boxed{\text{VARS}}$ $\boxed{\triangleright}$ 1 2 immediately after $\boxed{L_3}$ in the sequence of keystrokes above.

Press $\boxed{\text{ENTER}}$ to see the coefficients of the regression equation. We get $y = 0.775x + 5.05$.

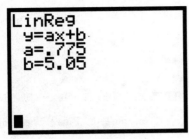 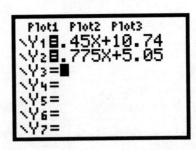

(e) To estimate the year in which the number of travelers to Europe will be the same as the number of travelers to Canada, we solve the system of equations

$$y = 0.45x + 10.74,$$
$$y = 0.775x + 5.05.$$

We graph the equations in the same viewing window and then use the Intersect feature to find their point of intersection. Through a trial-and-error process we find that $[0, 25, 0, 25]$, Xscl = 5, Yscl = 5, provides a good window in which to see this point.

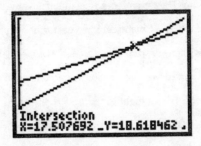

We see that the solution of the system of equations is approximately (17.51, 18.62), so the number of U. S. travelers to Europe will be the same as the number of U. S. travelers to Canada about 17.5 years after 1990, or in 2008.

ELIMINATION USING MATRICES

Matrices with up to 99 rows or columns can be entered on the TI-83 and TI-83 Plus. As many as ten matrices can be entered at one time. The row-equivalent operations necessary to write a matrix in row-echelon or reduced row-echelon form can be performed on the calculator, or we can go directly to reduced row-echelon form with a single command. We will illustrate the direct approach.

Section 8.6, Example 4 Solve the following system using a graphing calculator:

$$2x + 5y - 8z = 7,$$
$$3x + 4y - 3z = 8,$$
$$5y - 2x = 9.$$

First we rewrite the third equation in the form $ax + by + cz = d$:

$$2x + 5y - 8z = 7,$$
$$3x + 4y - 3z = 8,$$
$$-2x + 5y = 9.$$

Then we enter the coefficient matrix

$$\begin{bmatrix} 2 & 5 & -8 & 7 \\ 3 & 4 & -3 & 8 \\ -2 & 5 & 0 & 9 \end{bmatrix}$$

in the calculator. On the TI-83 press MATRX ▷ ▷ to display the MATRIX EDIT menu. On the TI-83 Plus press 2nd MATRX ▷ ▷. (MATRX is the second operation associated with the x^{-1} key on the TI-83 Plus.) Then select the matrix to be defined. We will select matrix [A] by pressing 1 or ENTER. Now the MATRIX EDIT screen appears. The dimensions of the matrix are displayed on the top line of this screen, with the cursor on the row dimension. Enter the dimensions of the coefficient matrix, 3 x 4, by pressing 3 ENTER 4 ENTER. Now the cursor moves to the element in the first row and first column of the matrix. Enter the elements of the first row by pressing 2 ENTER 5 ENTER (−) 8 ENTER 7 ENTER. The cursor moves to the element in the second row and first column of the matrix. Enter the elements of the second and third rows of the augmented matrix by typing each in turn followed by ENTER as above. Note that the screen only displays three columns of the matrix. The arrow keys can be used to move the cursor to any element at any time.

Matrix operations are found on the MATRIX MATH menu and are performed on the home screen. Press 2nd QUIT to leave the matrix editor and go to the home screen. Then, on the TI-83, press MATRX ▷ to access the MATRIX MATH menu. On the TI-83 Plus press 2nd MATRX ▷. The reduced row-echelon form command is item B on this menu. Copy it to the home screen by using the ▽ key to scroll down until "B" is highlighted and then press ENTER or by simply pressing ALPHA B. (We could also use the △ key to scroll up to "B" and then press ENTER.) We see the command "rref(" on the home screen.

 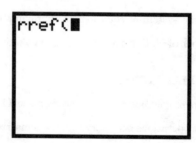

Since we want to find reduced row-echelon form for matrix [A], we enter [A] on the TI-83 by pressing MATRX ENTER) or MATRX 1). On the TI-83 Plus press 2nd MATRX ENTER) or 2nd MATRX 1). On either calculator we also press MATH 1 or MATH ENTER to see the elements of the reduced row-echelon form of the matrix in fraction form. Finally press ENTER to see the reduced row-echelon matrix. In the right-hand column we see that the solution of the system of equations is $\left(\frac{1}{2}, 2, \frac{1}{2}\right)$.

TI-83/83 Plus

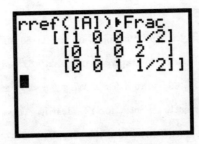

EVALUATING DETERMINANTS

We can evaluate determinants using the "det" operation from the MATRIX MATH menu.

Section 8.7, Example 3 Evaluate: $\begin{vmatrix} -1 & 0 & 1 \\ -5 & 1 & -1 \\ 4 & 8 & 1 \end{vmatrix}$.

First enter the 3 x 3 matrix

$$\begin{bmatrix} -1 & 0 & 1 \\ -5 & 1 & -1 \\ 4 & 8 & 1 \end{bmatrix}$$

as described on page 38 of this manual. We will enter it as matrix **A**.

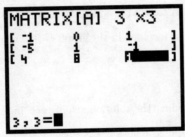

Then press $\boxed{\text{2nd}}$ $\boxed{\text{QUIT}}$ to go to the home screen. Next, on the TI-83, press $\boxed{\text{MATRX}}$ $\boxed{\triangleright}$ to access the MATRIX MATH menu. On the TI-83 Plus press $\boxed{\text{2nd}}$ $\boxed{\text{MATRX}}$ $\boxed{\triangleright}$. Press $\boxed{\text{ENTER}}$ to copy the "det(" operation to the home screen. Then enter the matrix name **A** on the TI-83 by pressing $\boxed{\text{MATRX}}$ $\boxed{\text{ENTER}}$ $\boxed{)}$ or $\boxed{\text{MATRX}}$ $\boxed{1}$ $\boxed{)}$. On the TI-83 Plus press $\boxed{\text{2nd}}$ $\boxed{\text{MATRX}}$ $\boxed{\text{ENTER}}$ $\boxed{)}$ or $\boxed{\text{2nd}}$ $\boxed{\text{MATRX}}$ $\boxed{1}$ $\boxed{)}$. Finally, press $\boxed{\text{ENTER}}$ to find the value of the determinant of matrix **A**.

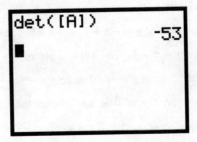

INEQUALITIES IN TWO VARIABLES

The solution set of an inequality in two variables can be graphed on the TI-83 and the TI-83 Plus.

Section 8.9, Example 4 Use a graphing calculator to graph the inequality $8x + 3y > 24$.

First we write the related equation, $8x + 3y = 24$, and solve it for y. We get $y = -\frac{8}{3}x + 8$. We will enter this as Y_1. Press $\boxed{Y=}$ to go to the equation-editor screen. If there is currently an entry for Y_1, clear it. Also clear or deselect any other equations that are entered. Now enter $y_1 = (-8/3)x + 8$. Since the inequality states that $8x + 3y > 24$, or y is *greater than* $-\frac{8}{3}x + 8$, we want to shade the half-plane above the graph of y_1. To do this, move the cursor to the GraphStyle icon to the left of Y_1 and press $\boxed{\text{ENTER}}$ repeatedly until the symbol indicating the "shade above" GraphStyle appears. If the "line" GraphStyle was previously selected, the "Shade above" icon will appear after $\boxed{\text{ENTER}}$ is pressed two times. (To shade below a line we would press $\boxed{\text{ENTER}}$ until the "shade below" GraphStyle symbol appears.) Then press $\boxed{\text{ZOOM}}$ 6 to see the graph of the inequality in the standard viewing window.

 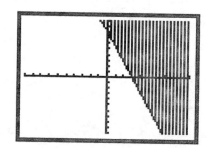

Note that when the "shade above" GraphStyle is selected it is not also possible to select the dotted GraphStyle so we must keep in mind the fact that the line $y = -\frac{8}{3}x + 8$ is not included in the graph of the inequality. If you graphed this inequality by hand, you would draw a dashed line.

SYSTEMS OF LINEAR INEQUALITIES

We can graph systems of inequalities by shading the solution set of each inequality in the system with a different pattern. When the "shade above" or "shade below" GraphStyle options are selected the calculator rotates through four shading patterns. Vertical lines shade the first function, horizontal lines the second, negatively sloping diagonal lines the third, and positively sloping diagonal lines the fourth. These patterns repeat if more than four functions are graphed.

Section 8.9, Example 8 Graph the system
$$x + y \leq 4,$$
$$x - y < 4.$$

First graph the equation $x + y = 4$, entering it in the form $y = -x + 4$. We determine that the solution set of $x + y \leq 4$ consists of all points on or below the line $x + y = 4$, or $y = -x + 4$, so we select the "shade below" GraphStyle for this function. Next graph $x - y = 4$, entering it in the form $y = x - 4$. The solution set of $x - y < 4$ is all points above the line $x - y = 4$, or $y = x - 4$, so for this function we choose the "shade above" GraphStyle. (See Example 4 above for instructions on selecting GraphStyle icons.) Now press $\boxed{\text{ZOOM}}$ 6 to display the solution sets of each inequality in the system and the region where they overlap in the standard viewing window. The region of overlap is the solution set of the system of inequalities. Keep in mind that the line $x + y = 4$, or $y = -x + 4$, is part of the solution set while $x - y = 4$, or $y = x - 4$, is not.

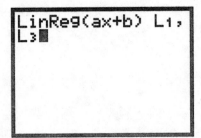 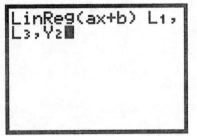

TI-83/83 Plus

Chapter 9
Exponents and Radical Functions

RADICAL EXPRESSIONS AND RATIONAL EXPONENTS

As discussed in Section 9.2, we can enter a radical expression using radical notation or rational exponents. For example, we can enter $y = \sqrt{x-3}$ using radical notation or as $y = (x-2)^{1/2}$ or as $y = (x-3)^{0.5}$. To enter $= \sqrt{x-3}$, press $\boxed{\text{2nd}}$ $\boxed{\sqrt{}}$ $\boxed{\text{X, T, }\Theta, n}$ $\boxed{-}$ $\boxed{3}$ $\boxed{)}$. ($\sqrt{}$ is the second operation associated with the $\boxed{x^2}$ key.) Note that the calculator supplies the left parenthesis along with the radical symbol and we add a right parenthesis after entering the radicand. To enter $y = (x-3)^{1/2}$, press $\boxed{(}$ $\boxed{\text{X, T, }\Theta, n}$ $\boxed{-}$ $\boxed{3}$ $\boxed{)}$ $\boxed{\wedge}$ $\boxed{(}$ $\boxed{1}$ $\boxed{\div}$ $\boxed{2}$ $\boxed{)}$. Note that both the radicand and the rational exponent are enclosed in parentheses. To enter $y = (x-3)^{0.5}$, press $\boxed{(}$ $\boxed{\text{X, T, }\Theta, n}$ $\boxed{-}$ $\boxed{3}$ $\boxed{)}$ $\boxed{\wedge}$ $\boxed{0}$ $\boxed{.}$ $\boxed{5}$. When the exponent is in decimal notation it is not necessary to enclose it in parentheses.

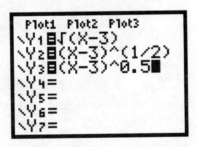

We can use either the cube root option or a rational exponent to enter a cube root. For example, to enter $y = \sqrt[3]{x+5}$ using radical notation we press $\boxed{\text{MATH}}$ $\boxed{4}$ $\boxed{\text{X, T, }\Theta, n}$ $\boxed{+}$ $\boxed{5}$ $\boxed{)}$. The keystrokes $\boxed{\text{MATH}}$ $\boxed{4}$ access the MATH MATH menu and then select item 4, the cube root option, from that menu. As with the square root option, the calculator supplies a left parenthesis and we close the parentheses after entering the radicand. Using a rational exponent, we can enter $y = \sqrt[3]{x+5}$ as $y = (x+5)^{1/3}$. Press $\boxed{(}$ $\boxed{\text{X, T, }\Theta, n}$ $\boxed{+}$ $\boxed{5}$ $\boxed{)}$ $\boxed{\wedge}$ $\boxed{(}$ $\boxed{1}$ $\boxed{\div}$ $\boxed{3}$ $\boxed{)}$. Since we cannot enter exact decimal notation for 1/3, we cannot use decimal notation for the exponent in this case.

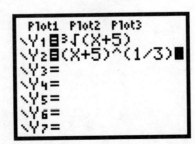

To enter $f(x) = \sqrt[4]{2x-7}$, as in Section 9.2, Example 3, we use the xth root option from the MATH MATH menu. To do this we first enter the index of the radical, 4. Then select the xth root feature and, finally, enter the radicand, $2x-7$. Press 4 $\boxed{\text{MATH}}$

5 $\boxed{(}$ 2 $\boxed{\text{X, T, }\Theta, n}$ $\boxed{-}$ 7 $\boxed{)}$. Note that the calculator does not supply a left parenthesis with this option, so we enter it ourselves. We could also enter this function as $f(x) = (2x-7)^{1/4}$ or as $f(x) = (2x-7)^{0.25}$.

Chapter 10
Quadratic Functions and Equations

MAXIMUMS AND MINIMUMS; FINDING THE VERTEX

We can use a graphing calculator to find the vertex of a quadratic function. We do this by using the Maximum or Minimum feature from the CALC menu.

Section 10.7, Example 4 Use a graphing calculator to determine the vertex of the graph of the function given by $f(x) = -2x^2 + 10x - 7$.

The coefficient of x^2 is negative, so we know that the graph of the function opens down and, thus, has a maximum value. Clear or deselect any functions previously entered on the equation-editor screen. Then enter $y = -2x^2 + 10x - 7$. Choose a viewing window that shows the vertex. One good choice is $[-3, 7, -10, 10]$.

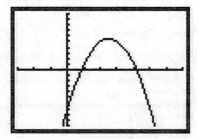

Now select the Maximum feature from the CALC menu by pressing $\boxed{\text{2nd}}$ $\boxed{\text{CALC}}$ 4. We are prompted to select a left bound for the vertex. Use the arrow keys to move the cursor to a point on the parabola to the left of the vertex or key in an x-value that is less than the x-coordinate of the vertex.

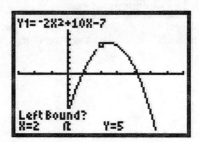

Press $\boxed{\text{ENTER}}$. Next we are prompted to select a right bound. Move the cursor to a point on the parabola to the right of the vertex or key in an x-value that is greater than the x-coordinate of the vertex.

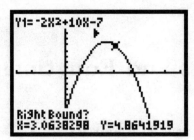

Press ENTER. We are now prompted to make a guess as to the x-coordinate of the vertex. Move the cursor close to the vertex or key in an x-value close to the x-value of the vertex.

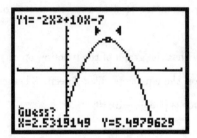

Press ENTER a third time. We see that the maximum function value is 5.5, and it occurs when x is approximately 2.5. Thus, the vertex of the graph of $f(x) = -2x^2 + 10x - 7$ is $(2.5, 5.5)$. (Note that, because of the method the calculator uses to find the maximum function value, the coordinates might not be exact and can vary slightly depending on the window chosen.)

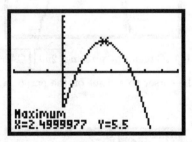

Minimum function values are found in a similar manner. Select the Minimum feature from the CALC menu by pressing 2nd CALC 3.

QUADRATIC REGRESSION

Regression can be used to fit a quadratic function to data when three or more data points are given.

Section 10.8, Example 4(c) According to the Centers for Disease Control and Prevention, the percent of high school students who reported having smoked a cigarette in the preceding 30 days declined from 1997 to 2001, after rising in the first part of the 1990s. Use the REGRESSION feature of a graphing calculator to fit a quadratic function $H(x)$ to all the given data in the following table.

Years after 1991	Percent of High School Students Who Smoked a Cigarette in the Preceding 30 Days
0	27.5
2	30.5
4	34.9
6	36.4
8	34.9
10	28.5

We enter the data in L_1 and L_2 as described on page 22 of this manual.

Then select QuadReg from the STAT CALC menu by pressing $\boxed{\text{STAT}}$ $\boxed{\text{CALC}}$ 5 $\boxed{\text{ENTER}}$. The calculator returns the coefficients of a quadratic function $y = ax^2 + bx + c$. From the screen below we see that we have $H(x) = -0.3151785714x^2 + 3.433214286x + 26.50714286$.

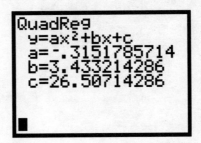

In order to use this function to perform computations it must be copied to the equation-editor screen. See pages 23 and 24 of this manual for the procedure to follow. The function can be evaluated using one of the methods on page 18.

Chapter 11
Exponential and Logarithmic Functions

COMPOSITE FUNCTIONS

For functions y_1 and y_2, when we enter $y_1(y_2)$ on a TI-83 or TI-83 Plus we are entering the composition $y_1 \circ y_2$. The composite functions found in Section 11.1, Example 2 are checked using tables on a graphing calculator. To check that $f \circ g = \sqrt{x-1}$ when $f(x) = \sqrt{x}$ and $g(x) = x - 1$, enter $y_1 = \sqrt{x}$, $y_2 = x - 1$, $y_3 = \sqrt{x-1}$, and $y_4 = y_1(y_2)$ on the equation-editor screen. We use the VARS Y-VARS menu to enter y_4. To do this, position the cursor beside $Y_4 =$ and press $\boxed{\text{VARS}}$ $\boxed{\triangleright}$ 1 1 $\boxed{(}$ $\boxed{\text{VARS}}$ $\boxed{\triangleright}$ 1 2 $\boxed{)}$. Then compare the values of y_3 and y_4 in a table. We show a table with TblStart = 1, ΔTbl = 0.5, and Indpnt and Depend both set on Auto. Use the $\boxed{\triangleright}$ key to scroll across the table to see the Y_3- and Y_4-columns.

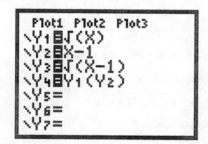

Similarly, to check that $g \circ f(x) = \sqrt{x} - 1$, also enter $y_5 = \sqrt{x} - 1$ and $y_6 = y_2(y_1)$. To enter y_6, position the cursor beside $Y_6 =$ and press $\boxed{\text{VARS}}$ $\boxed{\triangleright}$ 1 2 $\boxed{(}$ $\boxed{\text{VARS}}$ $\boxed{\triangleright}$ 1 1 $\boxed{)}$.

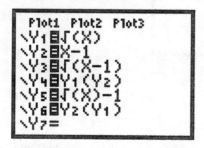

GRAPHING FUNCTIONS AND THEIR INVERSES

We can graph the inverse of a function using the DrawInv feature from the DRAW menu.

Section 11.1, Example 9(c) Graph the inverse of the function $g(x) = x^3 + 2$.

We will graph $g(x)$, $g^{-1}(x)$, and the line $y = x$ on the same screen. Press $\boxed{Y=}$ to go to the equation-editor screen and clear or deselect any existing entries. Then enter $y_1 = x^3 + 2$ and $y_2 = x$. Select a square window by pressing $\boxed{\text{ZOOM}}$ 5. Now paste the DrawInv command from the DRAW DRAW menu to the home screen by pressing $\boxed{\text{2nd}}$ $\boxed{\text{DRAW}}$ 8. Indicate that we want to draw the inverse of y_1 by pressing VARS ▷ 1 1. Finally press ENTER to see the graph of y_1^{-1} along with the graphs of y_1 and y_2. We show a window that has been squared from the standard window.

 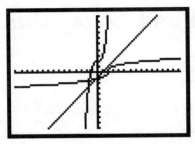

The drawing of y_1^{-1} can be cleared from the graph screen by pressing 2nd DRAW 1 to select the ClrDraw (clear drawing) operation. If ClrDraw was not accessed from the graph screen, it must be followed by ENTER. The graph will also be cleared when another function is subsequently entered on the "Y =" screen and graphed.

GRAPHING LOGARITHMIC FUNCTIONS

Section 11.3, Example 4 Graph: $f(x) = \log \dfrac{x}{5} + 1$.

We enter $y = \log(x/5) + 1$ on the equation-editor screen by positioning the cursor beside one of the function names and pressing LOG X, T, Θ, n ÷ 5) + 1. Note that the parentheses must be closed in the denominator of the logarithmic function. (Clear or deselect any previously entered functions.) We show the function graphed in the window $[-2, 10, -5, 5]$.

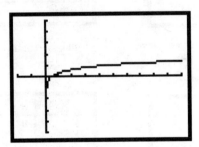

MORE ON GRAPHING

Section 11.5, Example 4 Graph: $f(x) = e^{-0.5x} + 1$.

We enter $y = e^{-0.5x} + 1$ on the equation-editor screen by positioning the cursor beside one of the function names and pressing 2nd e^x (−) . 5 X, T, Θ, n) + 1. (Clear or deselect any previously entered functions.) Select a window and press GRAPH. We show the function graphed in the window $[-5, 5, -2, 10]$.

TI-83/83 Plus

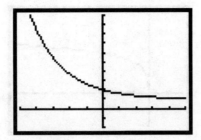

Section 11.5, Example 5(b) Graph: $f(x) = \ln(x+3)$.

We enter $y = \ln(x+3)$ on the equation-editor screen by positioning the cursor beside one of the function names and pressing $\boxed{\text{LN}}$ $\boxed{\text{X, T, }\Theta\text{, }n}$ $\boxed{+}$ $\boxed{3}$ $\boxed{)}$. (Clear or deselect any previously entered functions.) Select a window and press $\boxed{\text{GRAPH}}$. We show the function graphed in the window $[-5, 10, -5, 5]$.

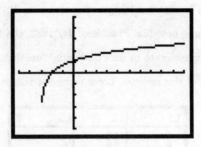

Section 11.5, Example 6 Graph: $f(x) = \log_7 x + 2$.

To use a graphing calculator we must first change the logarithmic base to e or 10. We will use e here. Recall that the change of base formula is $\log_b M = \dfrac{\log_a M}{\log_a b}$, where a and b are any logarithmic bases and M is any positive number. Let $a = e$, $b = 7$, and $M = x$ and substitute in the change-of-base formula. After clearing or deselecting previously entered functions, enter $y_1 = \dfrac{\ln x}{\ln 7} + 2$ on the equation-editor screen by positioning the cursor beside $Y_1 =$ and pressing $\boxed{\text{LN}}$ $\boxed{\text{X, T, }\Theta\text{, }n}$ $\boxed{)}$ $\boxed{\div}$ $\boxed{\text{LN}}$ $\boxed{7}$ $\boxed{)}$ $\boxed{+}$ 2. Note that the parentheses must be closed in both the numerator and the denominator.

Select a viewing window and press $\boxed{\text{GRAPH}}$. We show the graph in the window $[-2, 8, -2, 5]$.

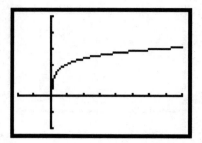

EXPONENTIAL REGRESSION

The STAT CALC menu contains an exponential regression feature.

Section 11.7, Example 9(a) In 1800, over 500,000 Tule elk inhabited the state of California. By the late 1800s, after the California Gold Rush, there were fewer than 50 elk remaining in the state. In 1978, wildlife biologists introduced a herd of 10 Tule elk into the Point Reyes National Seashore near San Francisco. By 1982, the herd had grown to 24 elk. There were 70 elk in 1986, 200 in 1996, and 500 in 2002. Use regression to fit an exponential function to the data and graph the function.

We enter the data as described on page 22 of this manual. Let x represent the number of years since 1978.

Now select ExpReg from the STAT CALC menu by pressing $\boxed{\text{STAT}}$ $\boxed{\triangleright}$ 0 $\boxed{\text{ENTER}}$ and also press $\boxed{\text{VARS}}$ $\boxed{\triangleright}$ 1 1 to copy the regression equation to the "Y =" screen. The calculator returns the values of a and b for the exponential function $y = ab^x$. We have $y = 13.01608148(1.168547698)^x$. We graph the equation in the window $[-2, 40, -5, 1000]$, Xscl = 5, Yscl = 100.

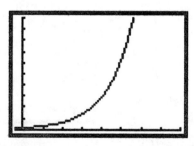

This function can be evaluated using one of the methods on page 18.

Chapter 12
Conic Sections

GRAPHING CIRCLES

Because the TI-83 and the TI-83 Plus can graph only functions, the equation of a circle must be solved for y before it can be entered in the calculator. Consider the circle $(x-3)^2 + (y+1)^2 = 16$ discussed in Section 12.1. In the text it is shown that this is equivalent to $y = -1 \pm \sqrt{16 - (x-3)^2}$. One way to graph this circle is first to enter $y_1 = -1 + \sqrt{16 - (x-3)^2}$ and $y_2 = -1 - \sqrt{16 - (x-3)^2}$. Then select a square window, to eliminate distortion, and press $\boxed{\text{GRAPH}}$. (See page 21 of this manual for a discussion on squaring the viewing window.) We show the graph in the window $[-3, 9, -5, 3]$.

 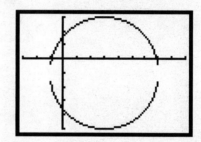

If the center and radius of a circle are known, the circle can be graphed using the Circle feature from the DRAW menu. Consider the circle $(x-3)^2 + (y+1)^2 = 16$ again.

The center of this circle is $(3, -1)$ and its radius is 4. To graph it using the Circle feature from the DRAW menu first press $\boxed{\text{Y =}}$ and clear all previously entered equations. Then select a square window. We will use $[-3, 9, -5, 3]$ as we did above. Press $\boxed{\text{2nd}}$ $\boxed{\text{QUIT}}$ to go to the home screen. Then press $\boxed{\text{2nd}}$ $\boxed{\text{DRAW}}$ 9 to display "Circle(." Enter the coordinates of the center and the radius, separating the entries by commas, and close the parentheses: 3 $\boxed{,}$ $\boxed{(-)}$ 1 $\boxed{,}$ 4 $\boxed{)}$ $\boxed{\text{ENTER}}$.

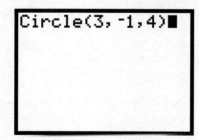 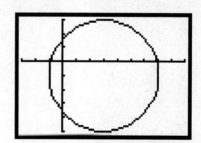

The drawing of the circle can be cleared from the graph screen by pressing $\boxed{\text{2nd}}$ $\boxed{\text{DRAW}}$ 1 to select the ClrDraw (clear drawing) operation. If ClrDraw is not accessed from the graph screen, it must be followed by $\boxed{\text{ENTER}}$. The graph will also be cleared when a function is subsequently entered on the "Y =" screen and graphed.

Chapter 13
Sequences, Series, and Probability

SEQUENCE MODE

To enter a sequence on a TI-83 or TI-83 Plus, first select Seq (Sequence) mode.

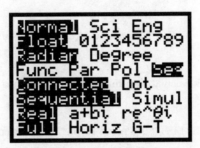

When the $\boxed{\text{X, T, }\Theta, n}$ key is pressed in Sequence mode, the variable that appears is n instead of x. In addition, the function names that appear on the equation-editor screen when $\boxed{Y=}$ is pressed are u, v, and w rather than Y_1, Y_2, and so on.

Section 13.1, Example 1 Find the first four terms and the 13th term of the sequence for which the general term is given by $a_n = (-1)^n n^2$.

After selecting Sequence mode, press $\boxed{Y=}$ to go to the sequence-editor screen. The minimum value of n in this sequence is 1, so we set nMin = 1. Then we enter the general term of the sequence beside "$u(n) =$" by pressing $\boxed{(}$ $\boxed{(-)}$ $\boxed{1}$ $\boxed{)}$ $\boxed{\wedge}$ $\boxed{\text{X, T, }\Theta, n}$ $\boxed{\times}$ $\boxed{\text{X, T, }\Theta, n}$ $\boxed{x^2}$.

Now set up a table with Indpnt set to Ask. (See pages 9 and 10 of this manual.) To see the first four terms and the 13th term of the sequence enter 1, 2, 3, 4, and 13 for n in the table.

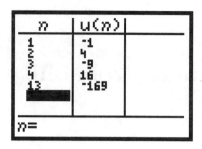

THE SEQUENCE FEATURE

The Sequence feature of the TI-83 and the TI-83 Plus writes the terms of a sequence as a list. This feature can be used even if the calculator is not in Sequence mode.

Section 13.1, Example 2 Use a graphing calculator to find the first five terms of the sequence for which the general term is given by $a_n = n/(n+1)^2$.

We will copy the Sequence feature from the LIST OPS menu to the home screen by pressing [2nd] [LIST] [▷] 5. (LIST is the second operation associated with the [STAT] key.) Now enter the general term of the sequence, the variable, and the values of the variable for the first and last terms we wish to calculate, all separated by commas. Press [X, T, Θ, n] [÷] [(] [X, T, Θ, n] [+] 1 [)] [x^2] [,] [X, T, Θ, n] [,] 1 [,] 5 [)]. We will also choose to display the terms of the sequence as fractions by pressing [MATH] 1 following the keystrokes shown above. Now press [ENTER] to see a list of the first five terms of the sequence. Note that we must use the [▷] key to see the fourth and fifth terms in the list.

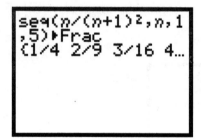
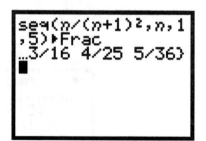

FINDING PARTIAL SUMS

We can use a graphing calculator to find partial sums of a sequence for which the general term is given by a formula.

Section 13.1, Example 5 Use a graphing calculator to find S_1, S_2, S_3, and S_4 for the sequence in which the general term is given by $a_n = (-1)^n/(n+1)$.

We will use the cumSum feature from the LIST OPS menu. This option lists the cumulative, or partial, sums for a sequence defined using the Sequence feature discussed above. First copy cumSum to the home screen by pressing [2nd] [LIST] [▷] 6. Next copy the Sequence feature by pressing [2nd] [LIST] [▷] 5. Now enter the general term of the sequence, the variable, and the first and last partial sums we wish to calculate, all separated by commas. We will also select the Fraction option from the MATH

MATH menu so that the partial sums will be displayed as fractions. Press $\boxed{(}$ $\boxed{(-)}$ 1 $\boxed{)}$ $\boxed{\wedge}$ $\boxed{\text{X, T, }\Theta, n}$ $\boxed{\div}$ $\boxed{(}$ $\boxed{\text{X, T, }\Theta, n}$ $\boxed{+}$ 1 $\boxed{)}$ $\boxed{,}$ $\boxed{\text{X, t, }\Theta, n}$ $\boxed{,}$ 1 $\boxed{,}$ 4 $\boxed{)}$ $\boxed{)}$ $\boxed{\text{MATH}}$ 1 $\boxed{\text{ENTER}}$. Note that we must use the $\boxed{\triangleright}$ key to see S_3 and S_4.

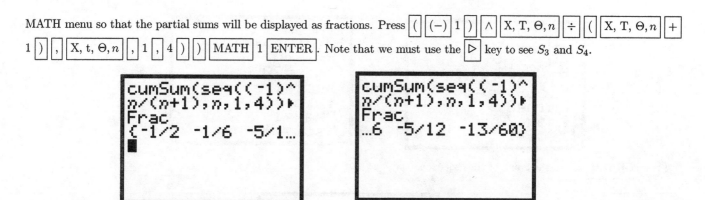

GRAPHS OF SEQUENCES

Section 13.1, Example 8 Graph the sequence for which the general term is given by $a_n = (-1)^n/n$.

The domain of a sequence is a set of integers, so the graph of a sequence is a set of points that are not connected. Thus, we use Dot mode to graph a sequence. Note that the calculator must also be set in Sequence mode.

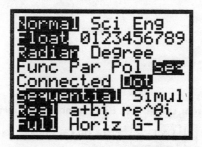

Press $\boxed{\text{Y}=}$ to go to the sequence-editor screen, and enter $u(n) = (-1)^n/n$ by positioning the cursor beside "$u(n) =$" and pressing $\boxed{(}$ $\boxed{(-)}$ 1 $\boxed{)}$ $\boxed{\wedge}$ $\boxed{\text{X, T, }\Theta, n}$ $\boxed{\div}$ $\boxed{\text{X, T, }\Theta, n}$. We also let nMin = 1.

Next we enter the window dimensions. We will graph the sequence from $n = 1$ through $n = 15$, so we let nMin = 1, nMax = 15, Xmin = 1, and Xmax = 20. A table of values of the sequence shows that the terms appear to be between -1 and 1, so we let Ymin = -1 and Ymax = 1 with Yscl = 0.1. We also set both PlotStart and PlotStep to 1. These settings cause the graph to begin with the first term in the sequence and to plot each term of the sequence.

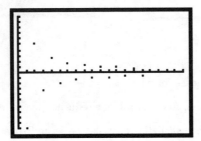

Press GRAPH to see the graph of the sequence.

EVALUATING FACTORIALS

Factorials can be evaluated on a graphing calculator.

Section 13.4, Example 3 Simplify: $\dfrac{8!}{5!3!}$.

We use the factorial feature, denoted !, from the MATH PRB (probability) menu. On the home screen press 8 MATH ◁ 4 ÷ (5 MATH ◁ 4 3 MATH ◁ 4) ENTER. Note that we must use parentheses in the denominator so that 8! is divided by both 5! and 3!. We could also access the MATH PRB menu by pressing MATH ▷ ▷ ▷ rather then MATH ◁, but we used the latter procedure since it requires fewer keystrokes.

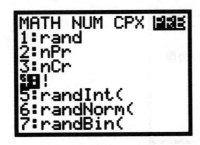
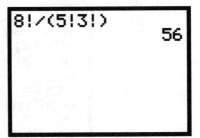

SIMPLIFYING $\binom{n}{r}$ NOTATION

Section 13.4, Example 4(a) Simplify: $\binom{7}{2}$.

The calculator uses the notation $_nC_r$ instead of $\binom{n}{r}$. This option is found in the MATH PRB menu. To simplify $\binom{7}{2}$, first press 7, then select option 3 from the MATH PRB menu by pressing MATH ◁ (or MATH ▷ ▷ ▷) 3, and then press 2 ENTER.

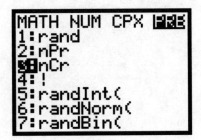 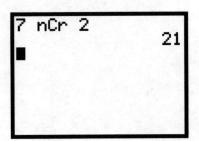

The TI-86 Graphics Calculator

Chapter 1
Introduction to Algebraic Expressions

GETTING STARTED

Press ON to turn on the TI-86 graphing calculator. (ON is the key at the bottom left-hand corner of the keypad.) You should see a blinking rectangle, or cursor, on the screen. If you do not see the cursor, try adjusting the display contrast. To do this, first press 2nd. (2nd is the yellow key in the left column of the keypad.) Then press and hold △ to increase the contrast or ▽ to decrease the contrast.

To turn the calculator off, press 2nd OFF. (OFF is the second operation associated with the ON key.) The calculator will turn itself off automatically after about five minutes without any activity.

Press 2nd MODE to display the MODE settings. (MODE is the second operation associated with the MORE key.) Initially you should select the settings on the left side of the display.

To change a setting on the Mode screen use ▽ or △ to move the blinking cursor to the line of that setting. Then use ▷ or ◁ to move the cursor to the desired setting and press ENTER. Press EXIT, CLEAR, or 2nd QUIT to leave the MODE screen. (QUIT is the second operation associated with the EXIT key.) Pressing EXIT, CLEAR, or 2nd QUIT will take you to the home screen where computations are performed.

It will be helpful to read the Introduction to the Graphing Calculator on pages 4 and 5 of the textbook as well as the Quick Start section and Chapter 1: Operating the TI-86 in your graphing calculator Guidebook before proceeding.

EVALUATING EXPRESSIONS

To evaluate expressions we substitute values for the variables.

Section 1.1, Example 4 Use a graphing calculator to evaluate $3xy + x$ for $x = 65$ and $y = 92$.

Enter the expression in the calculator, replacing x with 65 and y with 92. Press 3 × 6 5 × 9 2 + 6 5 ENTER. The calculator returns the value of the expression, 18,005.

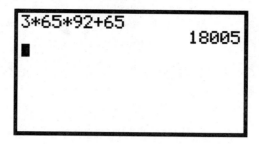

You can recall and edit your entry if necessary. If, for instance, in the expression above you pressed 8 instead of 9, first press 2nd ENTRY to return to the last entry. (ENTRY is the second operation associated with the ENTER key.) Then use the ◁ key to move the cursor to 8 and press 9 to overwrite it. If you forgot to type the 2, move the cursor to the plus sign; then press 2nd INS 2 to insert the 2 before the plus sign. (INS is the second operation associated with the DEL key.) You can continue to insert symbols immediately after the first insertion without pressing 2nd INS again. If you typed 31 instead of 3, move the cursor to 1 and press DEL. This will delete the 1. If you notice that an entry needs to be edited before you press ENTER to perform the computation, the editing can be done directly without recalling the entry.

The keystrokes 2nd ENTRY can be used repeatedly to recall entries preceding the last one. Pressing 2nd ENTRY twice, for example, will recall the next to last entry. Using these keystrokes a third time recalls the third to last entry and so on. The number of entries that can be recalled depends on the amount of storage they occupy in the calculator's memory.

USING A MENU

A menu is a list of options that appears when a key is pressed. Thus, multiple options, and sometimes multiple menus, may be accessed by pressing one key. For example, the following screen appears when 2nd MATH is pressed. (MATH is the second operation associated with the × multiplication key.) We see several submenus at the bottom of the screen. The F1 - F5 keys at the top of the keypad are used to select options from this menu. The arrow to the right of MISC indicates that there are more choices. They can be seen by pressing MORE.

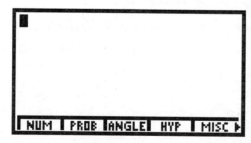

To choose the NUM submenu from the MATH menu press F1. (If you pressed MORE to see the additional items on the MATH menu as described above, now press MORE again to see the first five items on the menu. Then press F1 to choose NUM.) When NUM is chosen, the original submenus move up on the screen and the items on the NUM submenu appear at the bottom of the screen.

TI-86   65

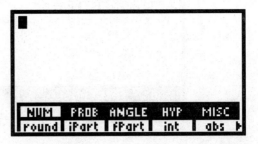

When two rows of options are displayed like this, the top row is accessed by pressing [2nd] followed by one of the keys [F1] - [F5]. These keystrokes access the second operations M1 - M5 associated with the [F1] - [F5] keys. The options on the bottom row are accessed by pressing one of the keys [F1] - [F5]. Absolute value, denoted "abs," is selected from the NUM submenu and copied to the home screen, for instance, by pressing [F5].

A menu can be removed from the screen by pressing [EXIT]. If both a menu and a submenu are displayed, press [EXIT] once to remove the submenu and twice to remove both.

The next example involves choosing an option from a menu.

Section 1.3, Example 12 Use a graphing calculator to find fraction notation for $\frac{2}{15} + \frac{7}{12}$.

Enter $\frac{2}{15} + \frac{7}{12}$ by pressing 2 [÷] 1 5 [+] 7 [÷] 1 2. Now press [ENTER] to find decimal notation for the sum. To convert this to fraction notation we recall the previous answer and then select the ▷Frac feature from the MISC submenu of the MATH menu. Press [2nd] [MATH] [F5] [MORE] [F1] [ENTER]. This will recall the previous answer and then convert it to fraction notation. Note that this conversion must be done immediately after the calculation is performed in order to have the result of the calculation available for the conversion.

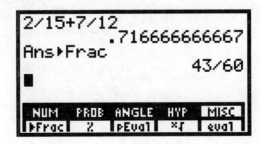

We can also find fraction notation for this sum in one step by selecting the ▷Frac operation before the sum is computed. When this is done, decimal notation does not appear on the screen. Press 2 [÷] 1 5 [+] 7 [÷] 1 2 [2nd] [MATH] [F5] [MORE] [F1] [ENTER].

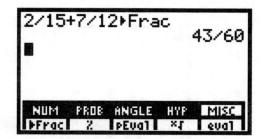

SQUARE ROOTS

Section 1.4, Example 5 Graph the real number $\sqrt{3}$ on a number line.

We can use the calculator to find a decimal approximation for $\sqrt{3}$. Press 2nd $\sqrt{}$ 3 ENTER.

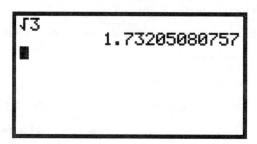

This approximation can now be used to locate $\sqrt{3}$ on the number line. The graph appears on page 33 of the text.

NEGATIVE NUMBERS AND ABSOLUTE VALUE

On the TI-86, $|x|$ is written abs(x). Recall that absolute value notation is found on the MATH NUM menu.

Section 1.3, Example 8 Find each absolute value: (a) $|-3|$; (b) $|7.2|$; (c) $|0|$.

When entering $|-3|$, keep in mind that the (−) key in the bottom row of the keypad must be used to enter a negative number on the calculator whereas the − key in the right-hand column of the keypad is used to enter subtraction. Press 2nd MATH F1 to access the MATH NUM menu. Then press F5 to copy "abs" to the home screen. Finally press (−) 3 ENTER to find $|-3|$. To find $|7.2|$ use the keystrokes above, replacing (−) 3 with 7 . 2 and to find $|0|$ replace (−) 3 with 0.

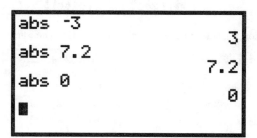

After finding $|-3|$, we could also find $|7.2|$ by recalling the previous entry ($|-3|$) and then editing that entry to replace -3 with 7.2. We could then recall the entry $|7.2|$ and edit it to find $|0|$. (See page 64 of this manual for a discussion of editing entries.)

Absolute value notation can also be found as the first item in the CATALOG and copied to the home screen. To do this first press [2nd] [CATLG-VARS] [F1] to access the catalog. (CATLG-VARS is the second operation associated with the [CUSTOM] key. A is the blue alphabetic operation associated with the [LOG] key.) Pressing A takes us to the first item in the Catalog that begins with A. A triangular selection cursor will be positioned beside the item "abs." Press [ENTER] to copy this item to the home screen. If the cursor is positioned beside "abs" after [F1] is pressed above, it is not necessary to press [A].

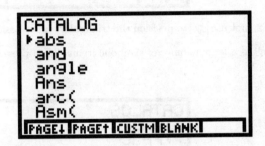

THE CUSTOM MENU

The TI-86 allows you to create a custom menu containing up to 15 items selected from the Catalog. To display the custom menu, press [CUSTOM].

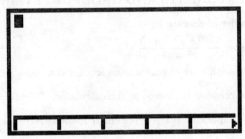

Press [MORE] once to see the second menu group and press [MORE] once again to see the third group.

To clear an item from the custom menu, press [2nd] [CATLG-VARS] [F1] [F4] to select BLANK from the Catalog menu. Then press one of the [F1] - [F5] keys corresponding to the location of the item to be cleared. To clear an item in the middle position of the first custom menu group, for instance, press [F3]. To clear an item in the second or third menu group, press [MORE] once or twice before pressing one of the [F1] - [F5] keys. A new item added to a custom menu will replace the item currently in that location, so it is not necessary to clear an item before one is added in its place.

We will put "abs" and "▷ Frac" in a custom menu to illustrate the procedure. To place "abs" in position F1 of the first menu group, first select Custom from the catalog menu be pressing [2nd] [CATLG-VARS] [F1] [F3]. Now move the triangular selection cursor in the Catalog to the first item that begins with A by pressing [A]. That item is "abs." Copy it to position F1 in the custom menu by pressing [F1].

To enter "▷ Frac" in position F2, first use $\boxed{\triangledown}$ to position the triangular cursor beside "▷ Frac" in the Catalog. This item follows the items beginning with Z as well as a large number of symbolic items. Then press $\boxed{F2}$ to copy the item to position F2.

ORDER OF OPERATIONS; EXPONENTS AND GROUPING SYMBOLS

The TI-86 follows the rules for order of operations.

Section 1.8, Example 8 Calculate: $\dfrac{12(9-7)+4\cdot 5}{2^4+3^2}$.

The fraction bar must be replaced with a set of parentheses around the entire numerator and another set of parentheses around the entire denominator when this expression is entered in the calculator. To enter an exponential expression, first enter the base, then use the $\boxed{\wedge}$ key followed by the exponent. If the exponent is 2, the keystrokes $\boxed{\wedge}$ 2 can be replaced with the single keystroke $\boxed{x^2}$.

To enter the expression above and express the result in fraction notation, first press $\boxed{(}$ 1 2 $\boxed{(}$ 9 $\boxed{-}$ 7 $\boxed{)}$ $\boxed{+}$ 4 $\boxed{\times}$ 5 $\boxed{)}$ $\boxed{\div}$ $\boxed{(}$ 2 $\boxed{\wedge}$ 4 $\boxed{+}$ 3 $\boxed{x^2}$ $\boxed{)}$. Remember to use the $\boxed{-}$ key for subtraction rather than the $\boxed{(-)}$ key, which is used to enter negative numbers. Then press $\boxed{\text{2nd}}$ $\boxed{\text{MATH}}$ $\boxed{F5}$ $\boxed{\text{MORE}}$ $\boxed{F1}$ $\boxed{\text{ENTER}}$ to choose the Frac operation from the MATH MISC menu. Finally, press $\boxed{\text{ENTER}}$ to see the result.

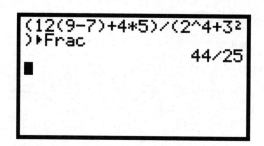

Chapter 2
Equations, Inequalities, and Problem Solving

EDITING ENTRIES

In **Section 2.1**, the procedure for editing entries is discussed on page 75. This procedure is described on page 64 of this manual.

EVALUATING FORMULAS; THE TABLE FEATURE: ASK MODE

Section 2.3, Example 2 Use the formula $B = 30a$, described in Example 1, to determine the minimum furnace output for well-insulated houses containing 800 ft^2, 1500 ft^2, 2400 ft^2, and 3600 ft^2.

First we replace B with y and a with x and enter the formula $y = 30x$ on the equation-editor screen as equation y_1. Press $\boxed{\text{GRAPH}}$ $\boxed{\text{F1}}$ to access this screen. If there is currently an expression displayed for y_1, clear it by positioning the cursor beside $y_1 =$ and pressing $\boxed{\text{CLEAR}}$. Do the same for expressions that appear on all other lines by using $\boxed{\triangledown}$ to move to a line and then pressing $\boxed{\text{CLEAR}}$. Note that any Stat Plots that had previously been turned on should be turned off. The names of the plots that are turned on are highlighted on the equation-editor screen. To turn off a plot, position the cursor over its name and press $\boxed{\text{ENTER}}$. Plots can also be turned off from the STAT menu. To turn off Plot 1, for example, press $\boxed{\text{2nd}}$ $\boxed{\text{STAT}}$ $\boxed{\text{F3}}$ $\boxed{\text{F5}}$ $\boxed{\text{ENTER}}$. Plot 1 can also be turned off by first pressing $\boxed{\text{2nd}}$ $\boxed{\text{STAT}}$ $\boxed{\text{F3}}$ $\boxed{\text{F1}}$ to go to the Plot 1 screen. Then position the cursor over Off and press $\boxed{\text{ENTER}}$.

To enter the equation, first use $\boxed{\triangle}$ or $\boxed{\triangledown}$ to move the cursor beside "$y1 =$." Now press 3 0 $\boxed{\text{x-VAR}}$ or 3 0 $\boxed{\text{F1}}$ to enter the right-hand side of the equation in the equation-editor screen. Note that the variable x can be entered either by pressing the $\boxed{\text{x-VAR}}$ key or by pressing $\boxed{\text{F1}}$ to select x from the $y(x) =$ submenu on the equation-editor screen.

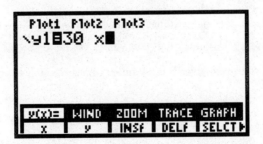

For an equation entered in the equation-editor screen, a table of x-and y-values can be displayed. We will use a table to evaluate the formula for the given values.

Once the equation in entered, press the $\boxed{\text{TABLE}}$ key in the second column of the keypad followed by $\boxed{\text{F2}}$ to access the TABLE SETUP screen. You can choose to supply the x-values yourself or you can set the calculator to supply them. To choose the x-values yourself, set "Indpnt" to "Ask" by positioning the cursor over "Ask" and pressing $\boxed{\text{ENTER}}$. In Ask mode the calculator disregards the settings of TblStart and ΔTbl.

Now press $\boxed{\text{F1}}$ to view the table. Values for x can be entered in the x-column of the table and the corresponding y-values will be displayed in the $y1$-column. To enter 800, 1500, 2400, and 3600, for instance, press 8 0 0 $\boxed{\text{ENTER}}$ 1 5 0 0 $\boxed{\text{ENTER}}$ 2 4 0 0 $\boxed{\text{ENTER}}$ 3 6 0 0 $\boxed{\text{ENTER}}$. The down arrow key $\boxed{\triangledown}$ can be pressed instead of $\boxed{\text{ENTER}}$ is desired.

We see that $y_1 = 24,000$ when $x = 800$, $y_1 = 45,000$ when $x = 1500$, $y_1 = 72,000$ when $x = 2400$, and $y_1 = 108,000$ when $x = 3600$, so the furnace outputs for 800 ft^2, 1500 ft^2, 2400 ft^2, and 3600 ft^2 are 24,000 Btu's, 45,000 Btu's, 72,000 Btu's, and 108,000 Btu's, respectively.

THE TABLE FEATURE: AUTO MODE

Section 2.4, Example 8 Village Stationers wants the customer service team to be able to look up the cost c of merchandise being returned when only the total amount paid T (including tax) is shown on the receipts. Use the formula $c = T/1.05$, developed in Example 7, to create a table of values showing cost given a total amount paid. Assume that the least possible sale is \$0.21.

We will use a graphing calculator to create the table of values. To enter the formula, we first replace c with y and T with x. Then we enter the formula on the equation-editor screen as $y = x/1.05$. Be sure that the Plots are turned off and that any previous entries are cleared. (See page 69 of this manual for the procedure for turning off the Plots and clearing equations.)

TI-86

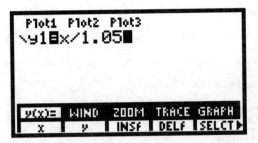

We will create a table in which the calculator supplies the x-values beginning with a value we specify and continuing by adding a value we specify to the preceding value for x. We will begin with an x-value of 0.21 (corresponding to $0.21) and choose successive increases of 0.01 (corresponding to $0.01). To do this first press TABLE F2 to access the TABLE SETUP screen. If "Indpnt" is set to "Auto," the calculator will supply values for x, beginning with the value specified as TblStart and continuing by adding the value of ΔTbl to the preceding value for x. Press . 2 1 ▽ . 0 1 to select a minimum x-value of 0.21 and an increment of 0.01. The "Indpnt" setting should be "Auto." If it is not, use the ▽ key to position the blinking cursor over "Auto" and then press ENTER . To display the table press F1 .

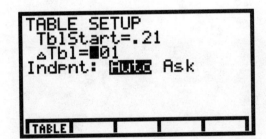

We can change the number of decimal places that the calculator will display to 2 so that the costs are rounded to the nearest cent. To do this, press 2nd MODE to access the MODE screen. Then press ▽ to move the cursor to the second line and press the ▷ key three times to highlight 2. (This assumes that "Float" was previously selected.) Finally, press ENTER to select two decimal places. To return to the table, press EXIT TABLE F1 .

Use the ▽ and △ keys to scroll through the table.

Before proceeding, return to the MODE screen and reselect "Float." This allows the number of decimal places to vary, or float, according to the computation being performed.

In **Section 2.5, Example 2**, two equations are entered on the equation-editor screen. The first equation, $y_1 = x + 1$, can be entered as described on page 69 of this manual. Then, to enter $y_2 = x + (x + 1)$ press either $\boxed{\triangledown}$ or $\boxed{\text{ENTER}}$ to position the cursor beside "$y2 =$." Now enter the right-hand side of the equation.

Chapter 3
Introduction to Graphing and Functions

If you selected 2 for the number of decimal places in Section 2.4, Example 8, and have not yet returned to the MODE screen to reselect "Float," do so now. In Float mode the number of decimal places varies, or floats, according to the computation being performed.

SETTING THE VIEWING WINDOW

The viewing window is the portion of the coordinate plane that appears on the graphing calculator's screen. It is defined by the minimum and maximum values of x and y: xMin, xMax, yMin, and yMax. The notation [xMin, xMax, yMin, yMax] is used in the text to represent these window settings or dimensions. For example, $[-12, 12, -8, 8]$ denotes a window that displays the portion of the x-axis from -12 to 12 and the portion of the y-axis from -8 to 8. In addition, the distance between tick marks on the axes is defined by the settings xScl and yScl. In this manual xScl and yScl will be assumed to be 1 unless noted otherwise. The setting xRes sets the pixel resolution. We usually select xRes = 1. The window corresponding to the settings $[-20, 30, -12, 20]$, xScl = 5, yScl = 2, xRes = 1, is shown below. Note that all of the entries on the equation-editor screen have been cleared in order to show this window. (See page 69 of this manual for the procedure for clearing entries and turning off the plots.)

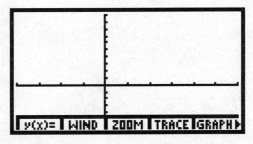

Press GRAPH F2 key to display the current window settings on your calculator in the WINDOW screen. The standard settings are shown below.

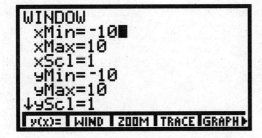

Section 3.1, Example 5 Set up a $[-100, 100, -5, 5]$ viewing window on a graphing calculator, choosing appropriate scales for the axes.

To change a setting, position the cursor beside the setting you wish to change and enter the new value. We will enter the given settings and let xScl = 10 and yScl =1. The choice of scales for the axes may vary. We select scaling that will allow some space between tick marks. If a scale that is too small is chosen, the tick marks will blend and blur. To change the settings to $[-100, 100, -5, 5]$, xScl = 10, yScl =1, press GRAPH F2 (−) 1 0 0 ENTER 1 0 0 ENTER 1 0 ENTER (−) 5 ENTER 5 ENTER 1 ENTER. The ▽ key may be used instead of ENTER after typing each window setting. To see the window with these settings, press F5.

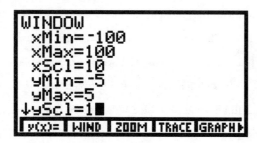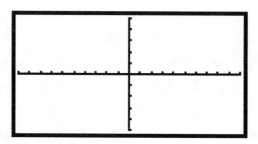

QUICK TIP: To return quickly to the standard window setting $[-10, 10, -10, 10]$, xScl = 1, yScl = 1, press GRAPH F3 F4.

GRAPHING EQUATIONS

After entering an equation and setting a viewing window, you can view the graph of an equation.

Section 3.2, Example 4 Graph $y = 2x$ using a graphing calculator.

Equations are entered on the equation-editor screen. Press GRAPH F1 to access this screen. If there is currently an expression displayed for y_1, clear it as described above. Do the same for expressions that appear on all other lines by using ▽ to move to a line and then pressing CLEAR. Note the any Stat Plots that had previously been turned on should be turned off. The names of the plots that are turned on are highlighted on the equation-editor screen. To turn off a plot, position the cursor over its name and press ENTER. Plots can also be turned off from the STAT menu as described above.

To enter the equation, first use △ or ▽ to move the cursor beside "$y1 =$." Now press 2 x-VAR or 2 F1 to enter the right-hand side of the equation in the equation-editor screen. Note that the variable x can be entered either by pressing the x-VAR key or by pressing F1 to select x from the $y(x) =$ submenu on the equation-editor screen.

The standard $[-10, 10, -10, 10]$ window is a good choice for this graph. Enter these dimensions in the WINDOW screen and then press F5 to see the graph or, from the equation-editor screen, simply press 2nd F3 F4 to select the standard window and see the graph. To remove the menu from the bottom of the screen, press CLEAR. The menu will reappear when GRAPH is pressed.

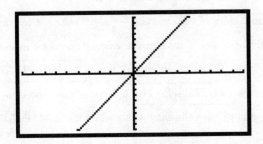

SOLVING EQUATIONS GRAPHICALLY: THE INTERSECT METHOD

We can use the Intersect feature from the GRAPH menu to solve equations.

Section 3.3, Example 2 Solve using a graphing calculator: $-\frac{3}{4}x + 6 = 2x - 1$.

On the equation editor screen, clear any existing entries and then enter $y_1 = -(3/4)x + 6$ and $y_2 = 2x - 1$. (Press ▽ or ENTER after entering y_1 to position the cursor beside "$y2 =$.") Although the parentheses in y_1 are not necessary, they make the equation easier to read on the equation-editor screen. Press 2nd M3 F4 to graph these equations in the standard viewing window. The solution of the equation $-\frac{3}{4}x + 6 = 2x - 1$ is the first coordinate of the point of intersection of these graphs. To use the Intersect feature to find this point, from the Graph screen first press MORE F1 MORE F3 to select ISECT from the GRAPH MATH menu. The query "First curve?" appears at the bottom of the screen. The blinking cursor is positioned on the graph of y_1. This is indicated by the number 1 in the upper right-hand corner of the screen. Press ENTER to indicate that this is the first curve involved in the intersection. Next the query "Second curve?" appears at the bottom of the screen. The blinking cursor is now positioned on the graph of y_2, indicated by the number 2 appearing in the top right-hand corner of the screen. Press ENTER to indicate that this is the second curve. We identify the curves since we could have as many as ten graphs on the screen at once. After we identify the second curve, the query "Guess?" appears at the bottom of the screen. Use the right and left arrow keys to move the blinking cursor close to the point of intersection of the graphs or type a number that approximates the first coordinate of the point of intersection. This provides the calculator with a guess as to the coordinates of this point. We do this since some pairs of curves can have more than one point of intersection. When the cursor is positioned or the approximation is typed, press ENTER a third time. Now the coordinates of the point of intersection appear at the bottom of the screen.

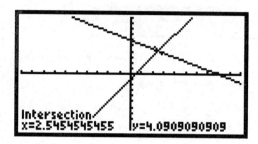

We see that $x = 2.5454545455$, so the solution of the equation is 2.5454545455.

We can check the solution by evaluating both sides of the equation $-\frac{3}{4}x + 6 = 2x - 1$ for this value of x. The first coordinate of the point of intersection has automatically been stored as x in the calculator, so we evaluate y_1 and y_2 for this value of x. First press 2nd QUIT to go to the home screen. Then to evaluate y_1 press 2nd CATLG-VARS MORE F4, position the triangular selection cursor beside "$y1$," and press ENTER ENTER. Repeat this procedure for y_2, positioning the triangular cursor beside "$y2$" this time. We see that y_1 and y_2 have the same value when x = 2.5454545455, so the solution checks.

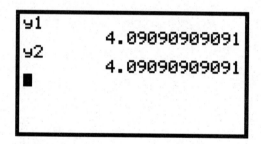

Note that although the procedure above verifies that 2.5454545455 is the solution, it is actually an approximation of the solution. In some cases the calculator will give an exact solution. Since the x-coordinate of the point of intersection is stored in the calculator as x, we can find an exact solution by converting x to fraction notation. This can be done from the home screen by pressing x-VAR 2nd MATH F5 MORE F1 ENTER. We see that the exact solution is 28/11.

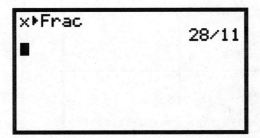

SOLVING EQUATIONS GRAPHICALLY: THE ZERO METHOD

As an alternative to the Intersect method, we can use the Root feature from the GRAPH MATH menu to find the zeros of a function. (Sometimes the words "zero" and "root" are used interchangeably.)

Section 3.3, Example 7 Solve $3 - 8x = 5 - 7x$ using the zero method.

First we get zero on one side of the equation:
$$3 - 8x = 5 - 7x$$
$$-2 - 8x = -7x$$
$$-2 - x = 0.$$

Then we graph $y = -2 - x$. We will use the standard window.

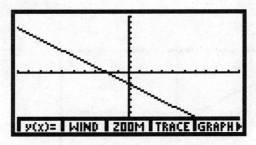

The first coordinate of the x-intercept of the graph is the zero of $y = -2 - x$ and thus is the solution of the original equation. From the GRAPH screen, press $\boxed{\text{MORE}}$ $\boxed{\text{F1}}$ $\boxed{\text{F1}}$ to select the Root feature from the GRAPH MATH menu. We are prompted to select a left bound. This means that we must choose an x-value that is to the left of the first coordinate of the x-intercept. This can be done by using the left and right arrow keys to move the cursor to a point on the graph that is to the left of the intercept or by keying in an x-value that is less than the first coordinate of the intercept.

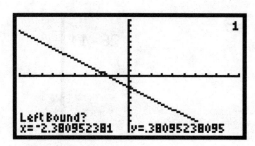

Once this is done, press ENTER. Now we are prompted to select a right bound that is to the right of the x-intercept. Again we can use the arrow keys or key in a value.

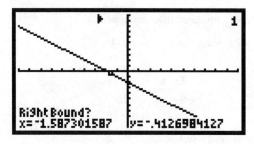

Press ENTER again. Finally, we are prompted to make a guess as to the value of the zero. Move the cursor to a point close to the zero or key in a value.

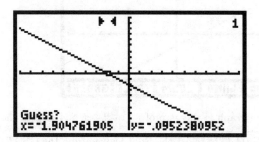

Press ENTER a third time. We see that $y = 0$ when $x = -2$, so -2 is the zero of $y = -2 - x$ and is thus the solution of the original equation, $3 - 8x = 5 - 7x$.

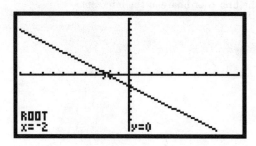

EVALUATING A FUNCTION

Function values can be found in several different ways on the TI-86.

Section 3.4, Example 5 For $f(a) = 2a^2 - 3a + 1$, find $f(3)$ and $f(-5.1)$.

One method for finding function values involves using function notation directly. To do this, first press `GRAPH` `F1` and enter the function on the equation-editor screen. Mentally replace a with x and $f(a)$ with y_1. Then enter $y_1 = 2x^2 - 3x + 1$. Now, to find $f(3)$, or $y_1(3)$, directly first press `2nd` `QUIT` to go to the home screen. Then press `2nd` `alpha` `Y` `1` `(` `3` `)` `ENTER`. We see that $y_1(3) = 10$, or $f(3) = 10$.

To find $f(-5.1)$, or $y_1(-5.1)$, we can repeat the previous procedure using -5.1 in place of 3, or we can edit the previous entry. To edit, copy the entry $y_1(3)$ to the home screen by pressing `2nd` `ENTRY`. Now replace 3 with -5.1 by first pressing `◁` `◁` to position the cursor over the 3. Then press `(-)` to overwrite the 3 with the negative symbol. To insert 5.1 after this symbol press `2nd` `INS` `5` `.` `1`. Finally press `ENTER` to find that $y_1(-5.1) = 68.32$, or $f(-5.1) = 68.32$.

```
y1(3)
                    10
y1(-5.1)
                 68.32
■
```

We can also find function values from the graph of the function.

Section 3.4, Example 6 Find $g(2)$ for $g(x) = 2x - 5$.

First press `GRAPH` `F1` to go to the equation editor screen and then clear any entries that are present. Also be sure that the Stat Plots are turned off. (See page 69 of this manual for instructions for clearing equations and turning off Stat Plots.) Now enter $y_1 = 2x - 5$ and press `2nd` `F3` `F4` to graph this function in the standard viewing window. We will use the EVAL feature from the GRAPH menu to find the value of y_1 when $x = 2$. This is $g(2)$. From the Graph window press `MORE` `MORE` `F1` to select EVAL. (If you pressed `CLEAR` to remove the menu from the Graph screen, press `GRAPH` first.) Now you must supply the value of x as indicated by the blinking cursor at the bottom of the screen beside "x =." Press 2 `ENTER`. We now see "x = 2, y = -1" at the bottom of the screen, so $g(2) = -1$.

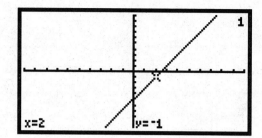

Chapter 4
Linear Equations, Inequalities, and Graphs

SQUARING THE VIEWING WINDOW

Section 4.4, Example 9 Determine whether the lines given by the equations $3x - y = 7$ and $x + 3y = 1$ are perpendicular, and check by graphing.

In the text each equation is solved for y in order to determine the slopes of the lines. We have $y = 3x - 7$ and $y = -\frac{1}{3}x + \frac{1}{3}$. Since $3\left(-\frac{1}{3}\right) = -1$, we know that the lines are perpendicular. To check this, we graph $y1 = 3x - 7$ and $y2 = -\frac{1}{3}x + \frac{1}{3}$. The graphs are shown below in the standard viewing window.

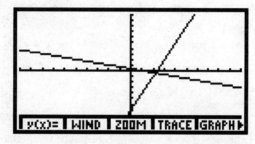

Note that the graphs do not appear to be perpendicular. This is due to the fact that, in the standard window, the distance between tick marks on the y-axis is about 3/5 the distance between tick marks on the x-axis. It is often desirable to choose window dimensions for which these distances are the same, creating a "square" window. On the TI-86, any window in which the ratio of the length of the y-axis to the length of the x-axis is 3/5 will produce this effect.

This can be accomplished by selecting dimensions for which yMax $-$ yMin $= \frac{3}{5}$(xMax $-$ xMin). For example, the windows $[-20, 20, -12, 12]$ and $[-10, 10, -6, 6]$ are square. When we change the dimensions to $[-10, 10, -6, 6]$ and press $\boxed{\text{GRAPH}}$ $\boxed{\text{F5}}$, the graphs now appear to be perpendicular as shown on the right below.

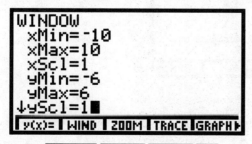 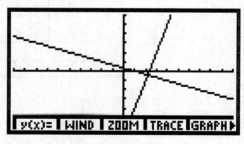

We could also press $\boxed{\text{GRAPH}}$ $\boxed{\text{ZOOM}}$ $\boxed{\text{MORE}}$ $\boxed{\text{F2}}$ and the calculator will select a square window.

ENTERING AND PLOTTING DATA; LINEAR REGRESSION

We can use the Linear Regression feature in the STAT CALC menu to fit a linear equation to a set of data.

Section 4.5, Example 9 The amount of paper recovered in the United States for various years is shown in the following table.

Years	Amount of Paper Recovered (in millions of tons)
1988	26.2
1990	29.1
1992	34.0
1994	39.7
1996	43.1
1998	45.1
2000	49.4

(a) Fit a linear function to the data.

(b) Graph the function and use it to estimate the amount of paper that will be recovered in 2003.

(a) We will enter the coordinates of the ordered pairs on the STAT list editor screen. To clear any existing lists first press 2nd STAT F2 (for EDIT). (STAT is the second operation associated with the + key.) Then use the arrow keys to move up to highlight "xStat" and press CLEAR ENTER. Do the same for yStat.

Once the lists are cleared, we can enter the coordinates of the points. We will enter the first coordinates (x-coordinates), as the number of years since 1988, in xStat and the second coordinates (y-coordinates), in millions of tons, in yStat. Position the cursor at the top of column xStat, below the xStat heading. To enter 0 (for 1988) press 0 ENTER. Continue typing the x-values 2, 4, 6, 8, 10, and 12, each followed by ENTER. The entries can be followed by ▽ rather than ENTER if desired. Press ▷ to move to the top of column yStat. Type the y-values 26.2, 29.1, 34.0, 39.7, 43.1, 45.1, and 49.4 in succession, each followed by ENTER or ▽. Note that the coordinates of each point must be in the same position in both lists.

Now press GRAPH F1 to go to the equation-editor screen and clear any equations that are currently entered. (See page 69 of this manual.) If you wish, instead of clearing an equation, you can deselect it. To do this, position the cursor beside the = sign and press F5. Note that the = sign is no longer highlighted, indicating that the equation has been deselected. The graph of an equation that has been deselected will not appear when the Graph screen is displayed. A deselected equation can be selected again by positioning the cursor beside the = sign and pressing F5. Note that the = sign is once again highlighted.

Now use the graphing calculator's linear regression feature to fit a linear equation to the data. Go to the home screen and press 2nd STAT F1 F3 to select linear regression, LinR, from the STAT CALC menu. Next enter the names of the lists that contain the variables x and y. Press 2nd LIST F3 F2 , F3 . Finally, press ENTER to see the coefficients of the regression equation $y = a + bx$. W see that the regression equation is $y = 1.97678571x + 26.225$.

Note that we also see the coefficient of correlation, denoted "corr." This number indicates how well the regression line fits the data. Press EXIT to remove the STAT CALC menu and display n, the number of data points. We have pressed EXIT twice to produce the screen below.

Immediately after the regression equation is found it can be copied to the equation-editor screen as $y1$. Note that any previous entry in $y1$ must have been cleared rather than deselected. Press GRAPH F1 and position the cursor beside $y1$. Then press 2nd CATLG-VARS MORE MORE F4 , use the ▽ key to position the cursor beside RegEq, and press ENTER . These keystrokes select the VARIABLES:STAT submenu from the CATLG-VARS menu, then select RegEq (Regression Equation) from this submenu, and paste it to the equation-editor screen.

Before the regression equation is found, it is possible to select a y-variable to which it will be stored on the equation-editor screen. After the data have been stored in the lists and the equation previously entered as $y1$ has been cleared, press 2nd STAT F1 F3 2nd LIST F3 F2 , F3 , 2nd alpha Y 1 ENTER . (Y is the blue alphabetic operation associated with the 0 numeric key.) The coefficients of the regression equation will be displayed on the home screen, and the regression equation will also be stored as $y1$ on the equation-editor screen.

(b) Now we will graph the regression equation. In order to see the data points along with the graph of the equation we will turn on and define a Stat Plot. To do this, press $\boxed{\text{2nd}}$ $\boxed{\text{STAT}}$ $\boxed{\text{F3}}$ to go to the STAT PLOT screen. Press $\boxed{\text{F1}}$ to select Plot 1 and then press $\boxed{\text{ENTER}}$ to turn on Plot 1. Next select the scatter diagram for Type, xStat for Xlist Name, yStat for Ylist Name, and the box for the Mark as shown below. To select Type, use the $\boxed{\triangledown}$ key to position the cursor beside Type = and then press one of the keys $\boxed{\text{F1}}$ - $\boxed{\text{F5}}$. Here we pressed $\boxed{\text{F1}}$ to select SCAT for a scatter diagram. Use the $\boxed{\triangledown}$ key again to position the cursor beside Xlist Name and press $\boxed{\text{F1}}$ to select xStat. Select yStat for Ylist Name similarly. The last item, Mark, allows us to choose a box, a cross, or a dot for each point. Here we have selected a box by positioning the cursor beside Mark = and pressing $\boxed{\text{F1}}$.

To select the dimensions of the viewing window notice that the years in the table range from 0 to 12 and the number of millions of tons of paper ranges from 26.2 to 49.4. We want to select dimensions that will include all of these values. One good choice is [0, 15, 0, 60], Yscl =10. Press $\boxed{\text{GRAPH}}$ $\boxed{\text{F2}}$ and enter these dimensions in the WINDOW screen.

Since the regression equation has been copied to the equation-editor screen, we can now press $\boxed{\text{GRAPH}}$ $\boxed{\text{F5}}$ to graph the regression line on the same axes as the data. Recall that, instead of entering window dimensions directly, from the Plot 1 screen we can press $\boxed{\text{GRAPH}}$ $\boxed{\text{F3}}$ $\boxed{\text{MORE}}$ $\boxed{\text{F5}}$ to activate the ZData operation which automatically selects a viewing window that contains all of the data points and also displays the graph.

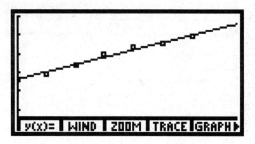

To estimate the amount of paper that will be recovered in 2003, evaluate the regression equation for $x = 15$. (2003 is 15 years after 1988.) The FCST (Forecast) feature from the STAT menu lends itself well to evaluating regression functions. Press 2nd STAT MORE F1. The blinking cursor appears beside $x =$. Enter 15 for x by pressing 1 5 ENTER. Now the cursor is positioned beside $y =$. Press F5 to see the value of y when $x = 15$. Instead of using the FCST feature, we could have used any of the methods for evaluating a function presented earlier in this chapter. (See page 79.) We see that when $x = 15, y \approx 55.9$, so we estimate that about 55.9 million tons of paper will be recovered in 2003.

THE TRACE FEATURE

Section 4.6, Example 10 Find the domain of $f + g$ if $f(x) = \sqrt{2x - 5}$ and $g(x) = \sqrt{x + 1}$.

In the text it is found that the domain of $f + g$ is $\left\{ x \middle| x \geq \dfrac{5}{2} \right\}$, or $\left[\dfrac{5}{2}, \infty \right)$. This can be confirmed, at least approximately, by using the TRACE feature to find function values on the graph of $f + g$. First graph $y = \sqrt{2x - 5} + \sqrt{x + 1}$. We will use the standard window. Now press F4 to select TRACE. Key in values for x and observe the corresponding y-values. We see that no y-values are given for x-values less than 2.5, or $\dfrac{5}{2}$. This indicates that x-values less than 2.5 are not in the domain of $f + g$. For x-values greater than or equal to 2.5, y-values are given and the ordered pairs appear to extend without bound, confirming that the domain of $f + g$ is $\left\{ x \middle| x \geq \dfrac{5}{2} \right\}$, or $\left[\dfrac{5}{2}, \infty \right)$.

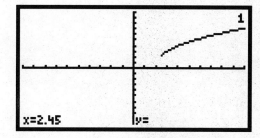 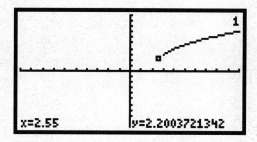

Chapter 5
Polynomials

CHECKING OPERATIONS ON POLYNOMIALS

A graphing calculator can be used to check operations on polynomials.

Section 5.3, Interactive Discovery Check the addition $(-3x^3 + 2x - 4) + (4x^3 + 3x^2 + 2) = x^3 + 3x^2 + 2x - 2$.

There are several ways in which we can use a graphing calculator to check this result. One of these is to compare the graphs of $y1 = (-3x^3 + 2x - 4) + (4x^3 + 3x^2 + 2)$ and $y2 = x^3 + 3x^2 + 2x - 2$. This is most easily done when different graph styles are used for the graphs.

Seven graph styles can be selected on the equation-editor screen of the TI-86. The **path graph style** can be used, along with the line style, to determine whether graphs coincide. To use graphs to check the addition in Example 9, first press GRAPH MORE F3 to determine whether Sequential graph format is selected. If it is not, position the blinking cursor over SeqG and then press ENTER. Next, on the equation-editor screen, enter $y1 = (-3x^3 + 2x - 4) + (4x^3 + 3x^2 + 2)$ and $y2 = x^3 + 3x^2 + 2x - 2$. We will select the line graph style for $y1$ and the path style for $y2$. To select these graph styles use ◁ to position the cursor anywhere in the equation and then press MORE. Now press F3 repeatedly until the desired style icon appears as shown on the right below.

The calculator will graph $y1$ first as a solid line. Then $y2$ will be graphed as the circular cursor traces the leading edge of the graph, allowing us to determine visually whether the graphs coincide. In this case, the graphs appear to coincide, so the factorization is probably correct.

We can also check the addition by **subtracting** the result from the original sum. With $y1$ and $y2$ entered as described above, position the cursor beside "$y3 =$" and use the CATLG-VARS menu to enter $y3 = y1 - y2$. First press 2nd CATLG-VARS MORE F4. Position the triangular cursor beside $y1$ and press ENTER. Then press − 2nd CATLG-VARS MORE F4. Now position the triangular cursor beside $y2$ and press ENTER.

If the subtraction is correct, $y1 = y2$, so $y3 = 0$. Since we are interested only in the values of $y3$, we deselect $y1$ and $y2$ as described on page 82 of this manual. Select the path style for $y3$ as described above.

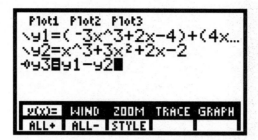

Now press [2nd] [F5] and determine if the graph of $y3$ is traced over the x-axis. Since it is, the sum is correct.

We can use a table of values to **compare values** of $y1$ and $y2$. If the expressions for $y1$ and $y2$ are the same for each given x-value, the result checks. If you deselected $y1$ and $y2$ to check the sum using subtraction as described above, select them again now. Then look at a table set in Auto mode. Since the values of $y1$ and $y2$ are the same for each given x-value, the result checks. Scrolling through the table to look at additional values makes this conclusion more certain.

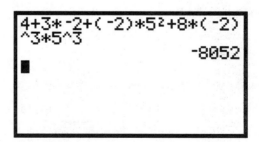

The TI-86 does not have a split screen feature that allows us to view a graph and a table of values simultaneously.

EVALUATING POLYNOMIALS IN SEVERAL VARIABLES

Section 5.6, Example 1 Evaluate the polynomial $4 + 3x + xy^2 + 8x^3y^3$ for $x = -2$ and $y = 5$.

To evaluate a polynomial in two or more variables, substitute numbers for the variables. This can be done by substituting values for x nd y directly or by storing the values of the variables in the calculator. The procedure for direct substitution is described on page 63 of this manual.

```
4+3*-2+(-2)*5²+8*(-2)
^3*5^3
                   -8052
```

To evaluate the polynomial by first storing -2 as x and 5 as y, we proceed as follows. To store -2 as x, press [(−)] 2 [STO▷] [x-VAR] [ENTER]. Now store 5 as y. Press 5 [STO▷] [2nd] [alpha] [Y] [ENTER]. alpha is the second operation associated with the blue ALPHA key in the left-hand column of the keypad, and Y is the alpha, or letter, operation associated with the 0 numeric

TI-86

key. Next enter the algebraic expression. Press 4 [+] 3 [x-VAR] [+] [x-VAR] [2nd] [alpha] [Y] [∧] 2 [+] 8 [x-VAR] [∧] 3 [2nd] [alpha] [Y] [∧] 3. Finally, press [ENTER] to find the value of the expression.

```
-2→x
             -2
5→y
              5
4+3 x+x*y²+8 x^3*y^3
          -8052
■
```

SCIENTIFIC NOTATION

To enter a number in scientific notation, first type the decimal portion of the number; then press [EE]; finally type the exponent, which can be at most two digits. For example, to enter 1.789×10^{-11} in scientific notation, press 1 [.] 7 8 9 [EE] [(-)] 1 1 [ENTER]. To enter 6.084×10^{23} in scientific notation, press 6 [.] 0 8 4 [EE] 2 3 [ENTER]. The decimal portion of each number appears before a small E while the exponent follows the E.

```
1.789E-11
            1.789E-11
6.084E23
             6.084E23
■
```

The graphing calculator can be used to perform computations in scientific notation.

Section 5.8, Example 9 Use a graphing calculator to check the computation $(7.2 \times 10^{-7}) \div (8.0 \times 10^6) = 9.0 \times 10^{-14}$.

We enter the computation in scientific notation. Press 7 [.] 2 [EE] [(-)] 7 [÷] 8 [EE] 6 [ENTER]. We have 9×10^{-14}, which checks.

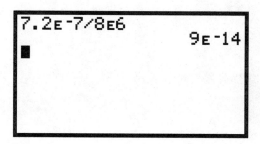

Chapter 6
Polynomials and Factoring

THE INTERSECT AND ZERO METHODS

Section 6.1, Example 1 Solve: $x^2 = 6x$.

To solve this equation by finding the x-coordinates of the points of intersection of $y_1 = x^2$ and $y_2 = 6x$, see the intersect method described on page 75 of this manual.

To solve this equation by writing it as $x^2 - 6x = 0$, graphing $f(x) = x^2 - 6x$, and then finding the values of x for which $f(x) = 0$, or the first coordinates of the x-intercepts of the graph, see the zero method described on page 77 of this manual.

POLYNOMIAL REGRESSION

The TI-86 has the capability to use regression to fit nonlinear polynomial equations to data.

Section 6.7, Example 4(a) The number of bachelor's degrees earned in the biological and life sciences for various years is shown in the following table. Fit a polynomial function of degree 3 (cubic) to the data.

Year	Number of Bachelor's Degrees Earned in Biological/Life Science
1971	35,743
1976	54,275
1980	46,370
1986	38,524
1990	37,204
1994	51,383
2000	63,532

First enter the data with the number of years since 1970 in xStat and the number of bachelor's degrees earned, in thousands, in yStat. (See page 82 of this manual for the procedure to follow.) We select cubic regression, denoted P3Reg, from the STAT CALC menu. Press [2nd] [QUIT] to go to the home screen. Then press [2nd] [STAT] [F1] [MORE] [F5] [2nd] [LIST] [F3] [F2] , [F3] , [2nd] [ALPHA] [Y] 1 [ENTER]. These keystrokes instruct the calculator to find the regression equation and to copy it to the equation-editor screen as $y1$. The calculator returns the number of data points along with the coefficients for a cubic function of the form $f(x) = ax^3 + bx^2 + cx + d$. Note that we must use the [▷] key to scroll across the screen to see all of the coefficients. Rounding the coefficients to the nearest thousandth, we have $f(x) = 0.009x^3 - 0.371x^2 + 4.176x + 34.415$.

We can use methods discussed earlier in this manual to estimate and predict function values and to find the year or years in which a specific function value occurs.

Chapter 7
Rational Expressions, Equations, and Functions

GRAPHING IN DOT MODE

Consider the graph of the function $T(t) = \dfrac{t^2 + 5t}{2t + 5}$ in Section 7.1, Example 1. Enter $y = (x^2 + 5x)/(2x + 5)$ and graph it in the window $[-5, 5, -5, 5]$.

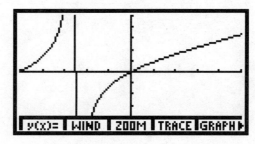

Note that a vertical line that is not part of the graph appears on the screen along with the two branches of the graph. The reason for this is discussed in the text.

This line will not appear if we change from DrawLine format to Dot format. Access the Format screen from the Graph screen by pressing MORE F3. Then move the cursor to DrawDot on the third line and press ENTER. Now press F5 to see the graph of the function in Dot format.

 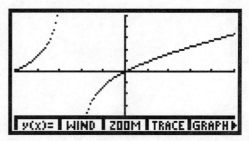

We can also select Dot format by selecting the "dot" Style on the equation-editor screen. If the function $T(t) = \dfrac{t^2 + 5t}{2t + 5}$ is entered as $y_1 = (x^2 + 5x)/(2x + 5)$, for instance, position the cursor beside $y_1 =$, press MORE, and then press F3 repeatedly until the "dot" icon appears. If the "line" icon was previously selected, F3 must be pressed six times to select the "dot" style.

Chapter 8
Systems of Equations and Problem Solving

SOLVING SYSTEMS OF EQUATIONS GRAPHICALLY

We can use the Intersect feature from the GRAPH MATH menu on the TI-86 to solve a system of two equations in two variables.

Section 8.1, Example 4(a) Solve graphically:

$$y - x = 1,$$
$$y + x = 3.$$

We graph the equations in the same viewing window and then find the coordinates of the point of intersection. Remember that equations must be entered in "$y =$" form on the equation-editor screen, so we solve both equations for y. We have $y = x + 1$ and $y = -x + 3$. Enter these equations, graph them in the standard viewing window, and find their point of intersection as described on page 75 of this manual. We see that the solution of the system of equations is (1, 2).

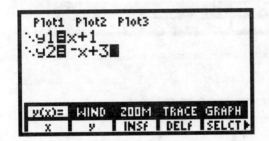 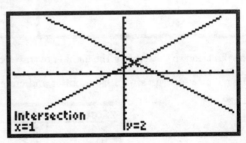

MODELS

Sometimes we model two situations with linear functions and then want to find the point of intersection of their graphs.

Section 8.1, Example 6 (d), (e) The numbers of U. S. travelers to Canada and to Europe are listed in the following table.

Year	U. S. Travelers to Canada (in millions)	U. S. Travelers to Europe (in millions)
1992	11.8	7.1
1994	12.5	8.2
1996	12.9	8.7
1998	14.9	11.1
2000	15.1	13.4

(d) Use linear regression to find two linear equations that can be used to estimate the number of U. S. travelers to Canada and Europe, in millions, x years after 1990.

(e) Use the equations found in part (d) to estimate the year in which the number of U. S. travelers to Europe will be the same as the number of U. S. travelers to Canada.

(d) To find the function $w(t)$, we enter the data for the years and waste generated in STAT lists as described on page 82 of this manual. We will express the years as the number of years after 1990 (in other words, 1990 is year 0) and enter them in xStat. Then enter the number of travelers to Canada, in millions, in yStat.

Now use linear regression to fit a linear function to the data. The function should also be copied to the equation editor screen. We will copy it as $y1$. See page 83 of this manual for the procedure to follow. We get $y1 = 10.74 + 0.45x$, or $y1 = 0.45x + 10.74$.

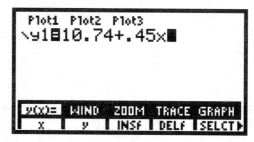

Next we fit a linear function to the data for the number of travelers to Europe, in millions, and save it as $y2$ on the equation-editor screen. We keep the number of years after 1990 in xStat and enter the number of travelers to Europe, in millions, in yStat.

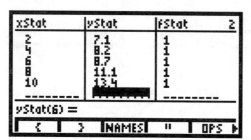

After entering the data, press EXIT to go to the home screen. Then press 2nd STAT F1 F3 2nd LIST F3 F2 , F3 , 2nd ALPHA Y 2 ENTER . We get $y2 = 5.05 + 0.775x$, or $y2 = 0.775x + 5.05$.

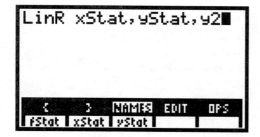

(e) To estimate the year in which the number of U. S. travelers to Europe will be the same as the number of U. S. travelers to Canada, we solve the system of equations

$$y = 0.45x + 10.74,$$
$$y = 0.775x + 5.05.$$

We graph the equations in the same viewing window and then use the Intersect feature to find their point of intersection. Through a trial-and-error process we find that $[0, 25, 0, 25]$ provides a good window in which to see this point.

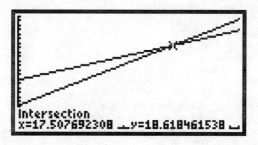

We see that the solution of the system of equations is approximately $(17.51, 18.62)$, so the number of U. S. travelers to Europe will be the same as the number of U. S. travelers to Canada about 17.51 yr after 1990, or in 2008.

ELIMINATION USING MATRICES

Matrices with up to 255 rows or columns can be entered on the TI-86. The row-equivalent operations necessary to write a matrix in row-echelon form or reduced row-echelon form can be performed on the calculator, or we can go directly to either of these forms with a single command. We will illustrate the direct approach for finding reduced row-echelon form.

Section 8.6, Example 4 Solve the following system using a graphing calculator:

$$2x + 5y - 8z = 7,$$
$$3x + 4y - 3z = 8,$$
$$5y - 2x = 9.$$

First we rewrite the third equation in the form $ax + by + cz = d$:

$$2x + 5y - 8z = 7,$$
$$3x + 4y - 3z = 8,$$
$$-2x + 5y = 9.$$

Then we enter the coefficient matrix

$$\begin{bmatrix} 2 & 5 & -8 & 7 \\ 3 & 4 & -3 & 8 \\ -2 & 5 & 0 & 9 \end{bmatrix}$$

in the calculator. Press $\boxed{\text{2nd}}$ $\boxed{\text{MATRX}}$ $\boxed{\text{F2}}$ to display the MATRIX EDIT screen. (MATRX is the second operation associated with the 7 numeric key.) The calculator is now in alphabetic mode, so we press $\boxed{\text{A}}$ to name the matrix "A." (A is the alphabetic operation associated with the $\boxed{\text{LOG}}$ key.) If other matrices have already been named, their names will appear at the bottom of the screen and can be chosen by pressing the $\boxed{\text{F1}}$ - $\boxed{\text{F5}}$ keys. Now press $\boxed{\text{ENTER}}$ to display the matrix editor. The dimensions of the matrix are displayed on the top line of this screen, with the cursor on the row dimension. Enter the dimensions of the coefficient matrix, 3 x 4, by pressing 3 $\boxed{\text{ENTER}}$ 4 $\boxed{\text{ENTER}}$. Now the cursor moves to the element in the first row and first column of the matrix. Enter the elements of the first row by pressing 2 $\boxed{\text{ENTER}}$ 5 $\boxed{\text{ENTER}}$ $\boxed{(-)}$ 8 $\boxed{\text{ENTER}}$ 7 $\boxed{\text{ENTER}}$. The cursor moves to the element in the second row and first column of the matrix. Enter the elements of the second and third rows of the augmented matrix by typing each in turn followed by $\boxed{\text{ENTER}}$ as above. Note that the screen only displays three columns of the matrix. The arrow keys can be used to move the cursor to any element at any time.

Matrix operations are found on the MATRIX OPS menu and are performed on the home screen. Press $\boxed{\text{2nd}}$ $\boxed{\text{QUIT}}$ leave the matrix editor and go to this screen. Then press $\boxed{\text{2nd}}$ $\boxed{\text{MATRX}}$ $\boxed{\text{F4}}$ to access the MATRIX OPS menu. The reduced row-echelon from command is item $\boxed{\text{F5}}$ "rref" on the menu. Copy it to the home screen by pressing $\boxed{\text{F5}}$. We see the command "rref."

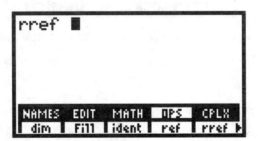

Since we want to find reduced row-echelon form for matrix A, we enter A by pressing $\boxed{\text{ALPHA}}$ $\boxed{\text{A}}$. Alternatively, press $\boxed{\text{2nd}}$ $\boxed{\text{F1}}$ to see the list of names. Then choose A by pressing the key $\boxed{\text{F1}}$ - $\boxed{\text{F5}}$ corresponding to A. To see the elements of the row-echelon form of the matrix in fraction form, press $\boxed{\text{2nd}}$ $\boxed{\text{MATH}}$ $\boxed{\text{F5}}$ $\boxed{\text{MORE}}$ $\boxed{\text{F1}}$. Finally press $\boxed{\text{ENTER}}$ to see this matrix. We see that the solution of the system of equations is $\left(\frac{1}{2}, 2, \frac{1}{2}\right)$.

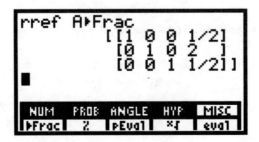

EVALUATING DETERMINANTS

We can evaluate determinants using the "det" operation from the MATRIX MATH menu.

Section 8.7, Example 3 Evaluate: $\begin{vmatrix} -1 & 0 & 1 \\ -5 & 1 & -1 \\ 4 & 8 & 1 \end{vmatrix}$.

First enter the 3 x 3 matrix

$$\begin{bmatrix} -1 & 0 & 1 \\ -5 & 1 & -1 \\ 4 & 8 & 1 \end{bmatrix}$$

as described on page 98 of this manual. We will enter it as matrix **A**.

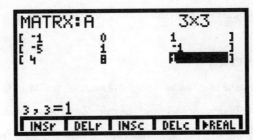

Then press $\boxed{\text{2nd}}$ $\boxed{\text{QUIT}}$ to go to the home screen. Next press $\boxed{\text{2nd}}$ $\boxed{\text{MATRX}}$ $\boxed{\text{F3}}$ to access the MATRIX MATH menu. Then press $\boxed{\text{F1}}$ to copy the "det" operation to the home screen. Next enter the matrix name **A** by pressing $\boxed{\text{2nd}}$ $\boxed{\text{F1}}$ $\boxed{\text{F1}}$. Finally, press $\boxed{\text{ENTER}}$ to find the value of the determinant of matrix **A**.

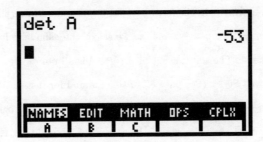

INEQUALITIES IN TWO VARIABLES

The solution set of an inequality in two variables can be graphed on the TI-86.

Section 8.9, Example 4 Use a graphing calculator to graph the inequality $8x + 3y > 24$.

First we write the related equation, $8x + 3y = 24$, and solve it for y. We get $y = -\frac{8}{3}x + 8$. We will enter this as $y1$. Press $\boxed{\text{GRAPH}}$ $\boxed{\text{F1}}$ to go to the equation-editor screen. If there is currently an entry for $y1$, clear it. Also clear or deselect any other equations that are entered and turn off the Plots. Now enter $y_1 = (-8/3)x + 8$. Since the inequality states that $8x + 3y > 24$, or y is *greater than* $-\frac{8}{3}x + 8$, we want to shade the half-plane above the graph of y_1. To do this, first press $\boxed{\text{MORE}}$. The choice "Style" appears above $\boxed{\text{F3}}$. Press $\boxed{\text{F3}}$ until the "shade above" Style icon appears to the left of $y1 =$. If the "line" Style was previously selected, the "Shade above" icon will appear after F3 is pressed two times. (To shade below a line we would press F3 until the "shade below" Style symbol appears.) Then press 2nd F3 F4 to see the graph of the inequality in the standard viewing window.

 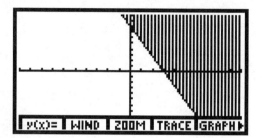

Note that when the "shade above" Style is selected it is not also possible to select the dotted Style so we must keep in mind the fact that the line $y = -\frac{8}{3}x + 8$ is not included in the graph of the inequality. If you graphed this inequality by hand, you would draw a dashed line.

SYSTEMS OF LINEAR INEQUALITIES

We can graph systems of inequalities by shading the solution set of each inequality in the system with a different pattern. When the "shade above" or "shade below" Style options are selected the calculator rotates through four shading patterns. Vertical lines shade the first function, horizontal lines the second, negatively sloping diagonal lines the third, and positively sloping diagonal lines the fourth. These patterns repeat if more than four functions are graphed.

Section 8.9, Example 8 Graph the system
$$x + y \leq 4,$$
$$x - y < 4.$$

First graph the equation $x + y = 4$, entering it in the form $y = -x + 4$. We determine that the solution set of $x + y \leq 4$ consists of all points on or below the line $x + y = 4$, or $y = -x + 4$, so we select the "shade below" Style for this function. Next graph $x - y = 4$, entering it in the form $y = x - 4$. The solution set of $x - y < 4$ is all points above the line $x - y = 4$, or $y = x - 4$, so for this function we choose the "shade above" Style. (See page 87 of this manual for instructions on selecting Style icons.) Now press 2nd F3 F4 to display the solution sets of each inequality in the system and the region where they overlap in the standard viewing window. The region of overlap is the solution set of the system of inequalities. Keep in mind that the line $x + y = 4$, or $y = -x + 4$, is part of the solution set while $x - y = 4$, or $y = x - 4$, is not.

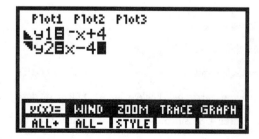 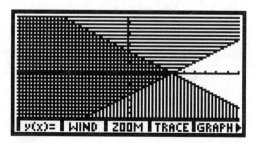

Chapter 9
Exponents and Radical Functions

RADICAL EXPRESSIONS AND RATIONAL EXPONENTS

As discussed in Section 9.2, we can enter a radical expression using radical notation or rational exponents. For example, we can enter $y = \sqrt{x-3}$ using radical notation or as $y = (x-2)^{1/2}$ or as $y = (x-3)^{0.5}$. To enter $= \sqrt{x-3}$, press $\boxed{\text{2nd}}$ $\boxed{\sqrt{}}$ $\boxed{(}$ $\boxed{\text{x-VAR}}$ $\boxed{-}$ 3 $\boxed{)}$. ($\sqrt{}$ is the second operation associated with the $\boxed{x^2}$ key.) Note that, since the radicand has more than one term, we must enclose it in parentheses. To enter $y = (x-3)^{1/2}$, press $\boxed{(}$ $\boxed{\text{x-VAR}}$ $\boxed{-}$ 3 $\boxed{)}$ $\boxed{\wedge}$ $\boxed{(}$ 1 $\boxed{\div}$ 2 $\boxed{)}$. Note that both the radicand and the rational exponent are enclosed in parentheses. To enter $y = (x-3)^{0.5}$, press $\boxed{(}$ $\boxed{X, T, \Theta, n}$ $\boxed{-}$ 3 $\boxed{)}$ $\boxed{\wedge}$ 0 $\boxed{.}$ 5. When the exponent is in decimal notation it is not necessary to enclose it in parentheses.

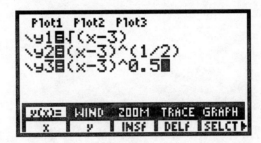

We can use either the xth root option from the MATH MISC menu or a rational exponent to enter a cube root. For example, to enter $y = \sqrt[3]{x+5}$ using radical notation first enter the index of the radical, 3. Then select the xth root option and finally enter the radicand enclosed in parentheses. To do this, press 3 $\boxed{\text{2nd}}$ $\boxed{\text{MATH}}$ $\boxed{\text{F5}}$ $\boxed{\text{MORE}}$ $\boxed{\text{F4}}$ $\boxed{(}$ $\boxed{\text{x-VAR}}$ $\boxed{+}$ 5 $\boxed{)}$. The keystrokes $\boxed{\text{2nd}}$ $\boxed{\text{MATH}}$ $\boxed{\text{F5}}$ $\boxed{\text{MORE}}$ $\boxed{\text{F4}}$ access the MATH MISC menu and then select the xth root option from that menu. Using a rational exponent, we can enter $y = \sqrt[3]{x+5}$ as $y = (x+5)^{1/3}$. Press $\boxed{(}$ $\boxed{\text{x-VAR}}$ $\boxed{+}$ 5 $\boxed{)}$ $\boxed{\wedge}$ $\boxed{(}$ 1 $\boxed{\div}$ 3 $\boxed{)}$. Since we cannot enter exact decimal notation for 1/3, we cannot use decimal notation for the exponent in this case.

To enter $f(x) = \sqrt[4]{2x-7}$, as in Section 9.2, Example 3, we also use the xth root option from the MATH MISC menu. To do this we first enter the index of the radical, 4. Then select the xth root feature and, finally, enter the radicand, $2x - 7$. Press 4 $\boxed{\text{2nd}}$ $\boxed{\text{MATH}}$ $\boxed{\text{F5}}$ $\boxed{\text{MORE}}$ $\boxed{\text{F4}}$ $\boxed{(}$ 2 $\boxed{X, T, \Theta, n}$ $\boxed{-}$ 7 $\boxed{)}$. We could also enter this function as $f(x) = (2x-7)^{1/4}$ or as

$f(x) = (2x-7)^{0.25}$.

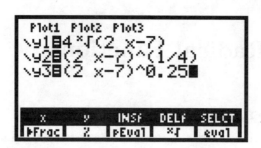

Chapter 10
Quadratic Functions and Equations

MAXIMUMS AND MINIMUMS; FINDING THE VERTEX

We can use a graphing calculator to find the vertex of a quadratic function. We do this by using the Maximum or Minimum feature from the GRAPH MATH menu.

Section 10.7, Example 4 Use a graphing calculator to determine the vertex of the graph of the function given by $f(x) = -2x^2 + 10x - 7$.

The coefficient of x^2 is negative, so we know that the graph of the function opens down and, thus, has a maximum value. Clear or deselect any functions previously entered on the equation-editor screen. Then enter $y = -2x^2 + 10x - 7$. Choose a viewing window that shows the vertex. One good choice is $[-3, 7, -10, 10]$.

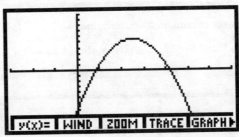

Now select the Maximum feature from the GRAPH MATH menu by pressing $\boxed{\text{MORE}}$ $\boxed{\text{F1}}$ $\boxed{\text{F5}}$. We are prompted to select a left bound for the vertex. Use the arrow keys to move the cursor to a point on the parabola to the left of the vertex or key in an x-value that is less than the x-coordinate of the vertex.

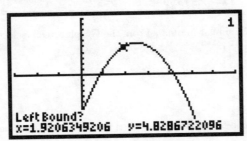

Press $\boxed{\text{ENTER}}$. Next we are prompted to select a right bound. Move the cursor to a point on the parabola to the right of the vertex or key in an x-value that is greater than the x-coordinate of the vertex.

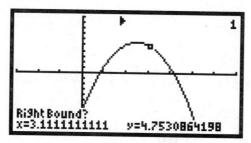

Press ENTER. We are now prompted to make a guess as to the x-coordinate of the vertex. Move the cursor close to the vertex or key in an x-value close to the x-value of the vertex.

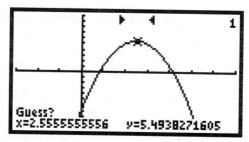

Press ENTER a third time. We see that the maximum function value is 5.5, and it occurs when x is approximately 2.5. Thus, the vertex of the graph of $f(x) = -2x^2 + 10x97$ is $(2.5, 5.5)$. (Note that, because of the method the calculator uses to find the maximum function value, the coordinates might not be exact and can vary slightly depending on the window chosen.)

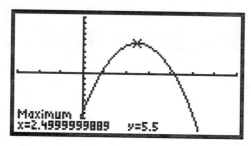

Minimum function values are found in a similar manner. From the Graph screen, select the Minimum feature from the GRAPH MATH menu by pressing MORE F1 F4.

QUADRATIC REGRESSION

Regression can be used to fit a quadratic function to data when three or more data points are given.

Section 10.8, Example 4(c) According to the Centers for Disease Control and Prevention, the percent of high school students who reported having smoked a cigarette in the preceding 30 days declined from 1997 to 2001, after rising in the first part of the 1990s. Use the REGRESSION feature of a graphing calculator to fit a quadratic function $H(x)$ to all the given data in the following table.

Years after 1991	Percent of High School Students Who Smoked a Cigarette in the Preceding 30 Days
0	27.5
2	30.5
4	34.9
6	36.4
8	34.9
10	28.5

We enter the data in xStat and yStat as described on page 82 of this manual.

Then press 2nd QUIT to go to the home screen and select P2Reg from the STAT CALC menu by pressing 2nd STAT F1 MORE F4 2nd LIST F3 F2 , F3 , 2nd alpha Y 1 ENTER . The calculator returns the coefficients of a quadratic function $y = ax^2 + bx + c$ and copies the function to the equation-editor screen as y1. We have $H(x) = -0.315178571429x^2 + 3.43321628571x + 26.5071428571$.

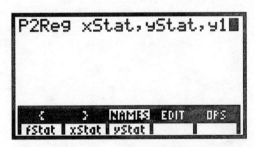

The function can be evaluated using one of the methods on pages 79 and 85 of this manual.

Chapter 11
Exponential and Logarithmic Functions

COMPOSITE FUNCTIONS

For functions y_1 and y_2, when we enter $y_1(y_2)$ on a graphing calculator we are entering the composition $y_1 \circ y_2$. The composite functions found in Section 11.1, Example 2 are checked using tables on a graphing calculator. To check that $f \circ g = \sqrt{x-1}$ when $f(x) = \sqrt{x}$ and $g(x) = x - 1$, enter $y_1 = \sqrt{x}$, $y_2 = x - 1$, $y_3 = \sqrt{x-1}$, and $y_4 = y_1(y_2)$ on the equation-editor screen. To enter y_4, position the cursor beside $y_4 =$ and press $\boxed{F2}$ $\boxed{1}$ $\boxed{(}$ $\boxed{F2}$ $\boxed{2}$ $\boxed{)}$. Then compare the values of y_3 and y_4 in a table. We show a table with TblStart $= 1$, ΔTbl $= 0.5$, and Indpnt and Depend both set on Auto. Use the $\boxed{\triangleright}$ key to scroll across the table to see the y_3- and y_4-columns.

Similarly, to check that $g \circ f(x) = \sqrt{x} - 1$, also enter $y_5 = \sqrt{x} - 1$ and $y_6 = y_2(y_1)$. To enter y_6, position the cursor beside $y_6 =$ and press $\boxed{F2}$ $\boxed{2}$ $\boxed{(}$ $\boxed{F2}$ $\boxed{1}$ $\boxed{)}$.

GRAPHING FUNCTIONS AND THEIR INVERSES

We can graph the inverse of a function using the DrInv feature from the DRAW menu.

Section 11.1, Example 9(c) Graph the inverse of the function $g(x) = x^3 + 2$.

We will graph $g(x)$, $g^{-1}(x)$, and the line $y = x$ on the same screen. Press $\boxed{\text{GRAPH}}$ $\boxed{F1}$ to go to the equation-editor screen and clear or deselect any existing entries. Then enter $y_1 = x^3 + 2$ and $y_2 = x$. Select a square window by pressing $\boxed{\text{2nd}}$ $\boxed{F3}$ $\boxed{\text{MORE}}$ $\boxed{F2}$. Now find the DrInv command in the Catalog by pressing $\boxed{\text{2nd}}$ $\boxed{\text{CATLG-VARS}}$ $\boxed{F1}$ $\boxed{D}$ and using the $\boxed{\triangledown}$ key to scroll down to DrInv. Copy this command to the home screen by pressing ENTER. Indicate that we want to draw the inverse of y_1 by pressing 2nd alpha Y 1. Finally press ENTER to see the graph of y_1^{-1} along with the graphs of y_1 and y_2. We show a window that has been squared from the standard window.

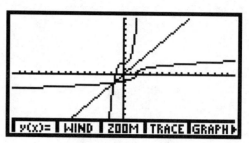

The drawing of y_1^{-1} can be cleared from the graph screen by pressing MORE F2 MORE MORE F1 to select the CLRDRW (clear drawing) operation. CLRDRW can also be accessed from the catalog. If this is done, ENTER must be pressed after the operation is pasted to the home screen.

GRAPHING LOGARITHMIC FUNCTIONS

Section 11.3, Example 4 Graph: $f(x) = \log \dfrac{x}{5} + 1$.

We enter $y = \log(x/5) + 1$ on the equation-editor screen by positioning the cursor beside one of the function names and pressing LOG (x-VAR ÷ 5) + 1. Note that the fraction must be entered in parentheses. (Clear or deselect any previously entered functions.) We show the function graphed in the window $[-2, 10, -5, 5]$.

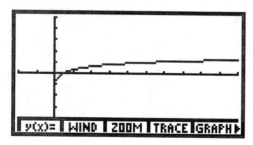

MORE ON GRAPHING

Section 11.5, Example 4 Graph: $f(x) = e^{-0.5x} + 1$.

We enter $y = e^{-0.5x} + 1$ on the equation-editor screen by positioning the cursor beside one of the function names and pressing 2nd e^x ((−) . 5 x-VAR) + 1. (Clear or deselect any previously entered functions.) Select a window and press F5. We show the function graphed in the window $[-5, 5, -2, 10]$.

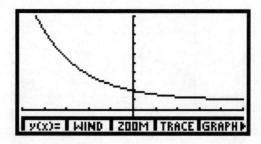

Section 11.5, Example 5(b) Graph: $f(x) = \ln(x+3)$.

We enter $y = \ln(x+3)$ on the equation-editor screen by positioning the cursor beside one of the function names and pressing $\boxed{\text{LN}}\;\boxed{(}\;\boxed{\text{x-VAR}}\;\boxed{+}\;\boxed{3}\;\boxed{)}$. (Clear or deselect any previously entered functions.) Select a window and, from the WINDOW screen, press $\boxed{\text{F5}}$. We show the function graphed in the window $[-5, 10, -5, 5]$.

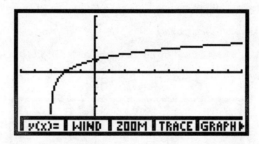

Section 11.5, Example 6 Graph: $f(x) = \log_7 x + 2$.

To use a graphing calculator we must first change the logarithmic base to e or 10. We will use e here. Recall that the change of base formula is $\log_b M = \dfrac{\log_a M}{\log_a b}$, where a and b are any logarithmic bases and M is any positive number. Let $a = e$, $b = 7$, and $M = x$ and substitute in the change-of-base formula. After clearing or deselecting previously entered functions, enter $y_1 = \dfrac{\ln x}{\ln 7} + 2$ on the equation-editor screen by positioning the cursor beside $y1 =$ and pressing $\boxed{\text{LN}}\;\boxed{\text{x-VAR}}\;\boxed{\div}\;\boxed{\text{LN}}\;\boxed{7}\;\boxed{+}\;2$.

Select a viewing window and press $\boxed{\text{F5}}$. We show the graph in the window $[-2, 8, -2, 5]$.

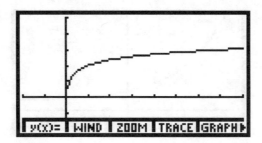

EXPONENTIAL REGRESSION

The STAT CALC menu contains an exponential regression feature.

Section 11.7, Example 9(a) In 1800, over 500,000 Tule elk inhabited the state of California. By the late 1800s, after the California Gold Rush, there were fewer than 50 elk remaining in the state. In 1978, wildlife biologists introduced a herd of 10 Tule elk into the Point Reyes National Seashore near San Francisco. By 1982, the herd had grown to 24 elk. There were 70 elk in 1986, 200 in 1996, and 500 in 2002. Use regression to fit an exponential function to the data and graph the function.

We enter the data as described on page 82 of this manual. Let x represent the number of years since 1978.

Now press 2nd QUIT to go to the home screen. Select ExpR from the STAT CALC menu by pressing 2nd F1 F5 2nd LIST F3 F2 , F3 ENTER . The calculator returns the values of a and b for the exponential function $y = ab^x$. We have $f(x) = 13.0160815(1.1685477)^x$.

This function can be copied to the equation-editor screen using one of the methods described on page 83 of this manual. It can be evaluated using one of the methods on pages 79 and 85.

Chapter 12
Conic Sections

GRAPHING CIRCLES

Because the TI-86 can graph only functions, the equation of a circle must be solved for y before it can be entered in the calculator. Consider the circle $(x-3)^2 + (y+1)^2 = 16$ discussed in Section 12.1. In the text it is shown that this is equivalent to $y = -1 \pm \sqrt{16 - (x-3)^2}$. One way to graph this circle is first to enter $y_1 = -1 + \sqrt{16 - (x-3)^2}$ and $y_2 = -1 - \sqrt{16 - (x-3)^2}$. Then select a square window, to eliminate distortion, and press $\boxed{\text{GRAPH}}$. (See page 81 of this manual for a discussion on squaring the viewing window.) We show the graph in the window $[-6, 14, -8, 4]$.

 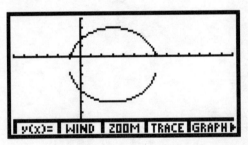

The software used to produce the graph above causes distortion. Nevertheless, when the circle is graphed on a graphing calculator in a square window, there is no distortion.

If the center and radius of a circle are known, the circle can be graphed using the Circle feature from the DRAW menu. Consider the circle $(x-3)^2 + (y+1)^2 = 16$ again.

The center of this circle is $(3, -1)$ and its radius is 4. To graph it using the Circle feature from the DRAW menu first press $\boxed{\text{GRAPH}}$ $\boxed{\text{F1}}$ and clear all previously entered equations. Then select a square window. We will use $[-6, 14, -8, 4]$ as we did above. Press $\boxed{\text{2nd}}$ $\boxed{\text{QUIT}}$ to go to the home screen. Then press $\boxed{\text{2nd}}$ $\boxed{\text{CATLG-VARS}}$ $\boxed{\text{F1}}$ $\boxed{\text{C}}$ $\boxed{\text{ENTER}}$ to copy "Circl(" to the home screen. Enter the coordinates of the center and the radius, separating the entries by commas, and close the parentheses: 3 $\boxed{,}$ $\boxed{(-)}$ 1 $\boxed{,}$ 4 $\boxed{)}$ $\boxed{\text{ENTER}}$.

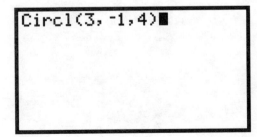 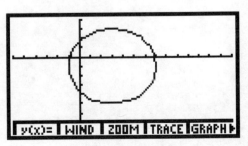

As mentioned above, the software used to produce the graph above causes distortion. Nevertheless, when the circle is graphed on a graphing calculator in a square window, there is no distortion.

The drawing of the circle can be cleared by first pressing $\boxed{\text{2nd}}$ $\boxed{\text{CATLG-VARS}}$ $\boxed{\text{F1}}$ to go to the catalog. Then position the cursor beside the entry "ClDrw" and press $\boxed{\text{ENTER}}$ to copy it to the home screen. Finally, press $\boxed{\text{ENTER}}$ to clear the drawing.

Chapter 13
Sequences, Series, and Probability

SEQUENCES

Section 13.1, Example 1 Find the first four terms and the 13th term of the sequence for which the general term is given by $a_n = (-1)^n n^2$.

Press $\boxed{\text{GRAPH}}$ $\boxed{\text{F1}}$ to go to the equation-editor screen. Then enter the general term of the sequence as $y1 = (-1)^x x^2$ by positioning the cursor beside $y1 =$ and pressing $\boxed{(}$ $\boxed{(-)}$ $\boxed{1}$ $\boxed{)}$ $\boxed{\wedge}$ $\boxed{\text{x-VAR}}$ $\boxed{\times}$ $\boxed{\text{x-VAR}}$ $\boxed{x^2}$. (Clear or deselect any other previously entered functions.)

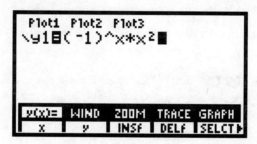

Now set up a table with Indpnt set to Ask. (See page 69 of this manual.) To see the first four terms and the 13th term of the sequence enter 1, 2, 3, 4, and 13 for n in the table.

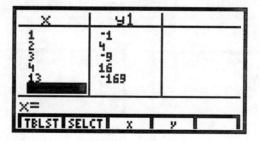

THE SEQUENCE FEATURE

The Sequence feature of the TI-86 writes the terms of a sequence as a list.

Section 13.1, Example 2 Use a graphing calculator to find the first five terms of the sequence for which the general term is given by $a_n = n/(n+1)^2$.

We will copy the Sequence feature from the LIST OPS menu to the home screen by pressing $\boxed{\text{2nd}}$ $\boxed{\text{LIST}}$ $\boxed{\text{F5}}$ $\boxed{\text{MORE}}$ $\boxed{\text{F3}}$. Now enter the general term of the sequence, the variable, and the values of the variable for the first and last terms we wish to calculate, all separated by commas. Press $\boxed{\text{x-VAR}}$ $\boxed{\div}$ $\boxed{(}$ $\boxed{\text{x-VAR}}$ $\boxed{+}$ $\boxed{1}$ $\boxed{)}$ $\boxed{x^2}$ $\boxed{,}$ $\boxed{\text{x-VAR}}$ $\boxed{,}$ $\boxed{1}$ $\boxed{,}$ $\boxed{5}$ $\boxed{)}$. We will also choose to display the terms of the sequence as fractions by pressing 2nd MATH F5 MORE F1 following the keystrokes shown above. Now press ENTER to see a list of the first five terms of the sequence. Note that we must use the ▷ key to see the fifth term in the list.

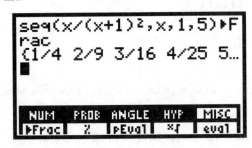
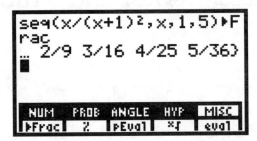

FINDING PARTIAL SUMS

We can use a graphing calculator to find partial sums of a sequence for which the general term is given by a formula.

Section 13.1, Example 5 Use a graphing calculator to find S_1, S_2, S_3, and S_4 for the sequence in which the general term is given by $a_n = (-1)^n/(n+1)$.

We will use the cSum feature from the LIST OPS menu. This option lists the cumulative, or partial, sums for a sequence defined using the Sequence feature discussed above. First copy cSum to the home screen by pressing 2nd LIST F5 MORE MORE F3. Next copy the Sequence feature by pressing 2nd LIST F5 MORE F3. Now enter the general term of the sequence, the variable, and the first and last partial sums we wish to calculate, all separated by commas. We will also select the Fraction option from the MATH MISC menu so that the partial sums will be displayed as fractions. Press ((−) 1) ∧ x-VAR ÷ (x-VAR + 1) , x-VAR , 1 , 4)) 2nd MATH F5 MORE F1 ENTER. Note that we must use the ▷ key to see S_4.

GRAPHING SEQUENCES

The TI-86 does not have a Sequence mode that allows us to graph sequences straightforwardly.

TI-86

EVALUATING FACTORIALS

Factorials can be evaluated on a graphing calculator.

Section 13.4, Example 3 Simplify: $\dfrac{8!}{5!3!}$.

We use the factorial feature, denoted !, from the MATH PROB (probability) menu. On the home screen press 8 [2nd] [MATH] [F2] [F1] [÷] [(] 5 [F1] 3 [F1] [)] [ENTER]. Note that we must use parentheses in the denominator so that 8! is divided by both 5! and 3!.

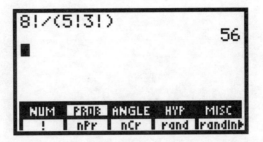

SIMPLIFYING $\binom{n}{r}$ NOTATION

Section 13.4, Example 4(a) Simplify: $\binom{7}{2}$.

The calculator uses the notation $_nC_r$ instead of $\binom{n}{r}$. This option is found in the MATH PROB menu. To simplify $\binom{7}{2}$, first press 7, then select $_nC_r$ from the MATH PROB menu by pressing [2nd] [MATH] [F2] [F3], and then press 2 [ENTER].

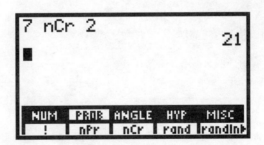

The TI-89 Graphics Calculator

Chapter 1
Introduction to Algebraic Expressions

GETTING STARTED

Press $\boxed{\text{ON}}$ to turn on the TI-89 graphing calculator. ($\boxed{\text{ON}}$ is the key at the bottom left-hand corner of the keypad.) The home screen is displayed. You should see a row of boxes at the top of the screen and two horizontal lines with lettering below them at the bottom of the screen. If you do not see anything, try adjusting the display contrast. To do this, first press $\boxed{\diamond}$. ($\boxed{\diamond}$ is the key in the left column of the keypad with a green diamond inside a green border. All operations associated with the $\boxed{\diamond}$ key are printed on the keyboard in green, the same color as the $\boxed{\diamond}$ key.) Then press $\boxed{+}$ to darken the display or $\boxed{-}$ to lighten the display. Be sure to use the black $\boxed{-}$ key in the right column of the keypad rather than the gray $\boxed{(-)}$ key on the bottom row.

One way to turn the calculator off is to press $\boxed{\text{2nd}}$ $\boxed{\text{OFF}}$. (OFF is the second operation associated with the $\boxed{\text{ON}}$ key. All operations accessed by using the $\boxed{\text{2nd}}$ key are printed on the keyboard in yellow, the same color as the $\boxed{\text{2nd}}$ key.) When you turn the TI-89 on again the home screen will be displayed regardless of the screen that was displayed when the calculator was turned off. $\boxed{\text{2nd}}$ $\boxed{\text{OFF}}$ cannot be used to turn off the calculator if an error message is displayed. The calculator can also be turned off by pressing $\boxed{\diamond}$ $\boxed{\text{OFF}}$. This will work even if an error message is displayed. When the TI-89 is turned on again the display will be exactly as it was when it was turned off. The calculator will turn itself off automatically after several minutes without any activity. When this happens the display will be just as you left it when you turn the calculator on again.

From top to bottom, the home screen consists of the toolbar, the large history area where entries and their corresponding results are displayed, the entry line where expressions or instructions are entered, and the status line which shows the current state of the calculator. These areas will be discussed in more detail as the need arises.

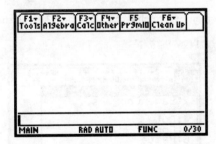

Press $\boxed{\text{MODE}}$ to display the MODE settings. Modes that are not currently valid, due to the existing choices of settings, are dimmed. Initially you should select the settings shown below.

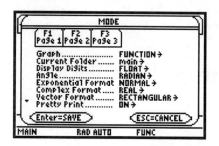
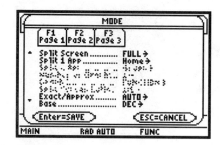
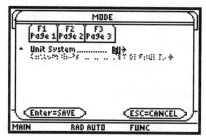

To change a setting on the Mode screen use ▽ or △ to move the cursor to the line of that setting. Then use ▷ to display the options. Press the number of the desired option followed by ENTER. Press HOME or 2nd QUIT to leave the MODE screen and return to the home screen. (QUIT is the second operation associated with the ESC key.) Note that the cursor skips dimmed settings as you move through the options.

It will be helpful to read the Introduction to the Graphing Calculator on pages 4 and 5 of the textbook as well as Chapter 1: Getting Started and Chapter 2: Operating the TI-89 in the TI-89 Guidebook before proceeding.

EVALUATING EXPRESSIONS

To evaluate expressions we substitute values for the variables.

Section 1.1, Example 4 Use a graphing calculator to evaluate $3xy + x$ for $x = 65$ and $y = 92$.

You might want to clear any previously entered computations from the history area of the home screen first. To do this, access Tools from the tool bar at the top of the screen by pressing F1, the blue key at the top left-hand corner of the keypad. Then select item 8, Clear Home, from this menu by pressing 8.

Then enter the expression in the calculator, replacing x with 65 and y with 92. Press 3 × 6 5 × 9 2 + 6 5 ENTER. The calculator returns the value of the expression, 18,005.

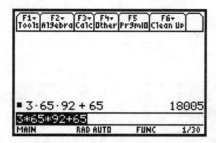

You can edit your entry if necessary. After ENTER is pressed to evaluate an expression, the TI-89 leaves the expression on the entry line and highlights it. To edit the expression you must first remove the highlight to avoid the possibility of accidently typing over the entire expression. To do this, press ◁ or ▷ to move the cursor (a blinking vertical line) toward the side of the expression to be edited. If, for instance, in the expression above you pressed 8 instead of 9, first press ◁ to move the cursor toward the 8. Now, to type a 9 over the 8, first select overtype mode by pressing 2nd INS. (INS is the second operation associated with the ← key.) Now the cursor becomes a dark, blinking rectangle rather than a vertical line. Use ▷ to position the cursor over the 8 and then press 9 to write a 9 over the 8. To leave overtype mode press 2nd INS again. The calculator is now in the insert mode, indicated by a vertical cursor, and will remain in that mode until overtype mode is once again selected.

If you forgot to type the 2, move the insert cursor to the left of the plus sign and press 2 to insert the parenthesis before the plus sign. You can continue to insert symbols immediately after the first insertion. If you typed 31 instead of 3, move the cursor to the left of 1 and press ←. This will delete the 1. Instead of using overtype mode to overtype a character as described above, we can use ← to delete the character and then, in insert mode, insert a new character.

If you accidently press △ instead of ◁ or ▷ while editing an expression, the cursor will move up into the history area of the screen. Press ESC to return immediately to the entry line. The ▽ key can also be used to return to the entry line. It must be pressed the same number of times the △ key was pressed accidently.

If you notice that an entry needs to be edited before you press ENTER to perform the computation, the editing can be done as described above without the necessity of first removing the highlight from the entry.

The keystrokes 2nd ENTRY can be used repeatedly to recall entries preceding the last one. (ENTRY is the second function associated with the ENTER key.) Pressing 2nd ENTRY twice, for example, will recall the next to last entry. Using these keystrokes a third time recalls the third to last entry and so on. The number of entries that can be recalled depends on the amount of storage they occupy in the calculator's memory.

Previous entries and results of computations can also be copied to the entry line by first using the △ key to move through the history area until the desired entry or result is highlighted. Then press ENTER to copy it to the entry line.

USING A MENU

In the previous example we used the Tools menu to clear the home screen. In general, a menu is a list of options that appear when a key is pressed. For example, we pressed F1 to display the Tools menu. We can select an item from a menu by using ▽ to highlight it and then pressing ENTER or by simply pressing the number of the item. If an item is identified by a letter rather than a number, press the purple [alpha] key followed by the letter of the item. The letters are printed in purple above the keys on the keypad. The down-arrow beside item 8 in the menu above indicates that there are additional items in the menu. Use [▽] to scroll down to them.

FRACTION NOTATION

Section 1.3, Example 12 Use a graphing calculator to find fraction notation for $\frac{2}{15} + \frac{7}{12}$.

Go to the entry line of the home screen and press 2 [÷] 1 5 [+] 7 [÷] 1 2 [ENTER].

Note that the result is expressed in fraction notation. This will occur when Exact is selected as the Exact/Approx mode on the Mode screen. It will also occur when the Auto setting is selected and there is no decimal point in the entry.

To see the result expressed in decimal notation, we can enter the expression as above and then press [◇] before pressing [ENTER]. Note the black ◇ at the bottom of the screen on the left below and the decimal result on the right.

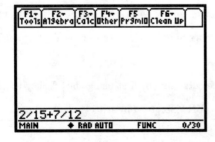 

We will also see the result expressed in decimal notation if Approximate is selected as the Exact/Approx mode before entering the expression.

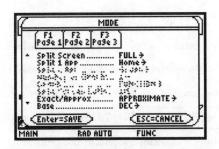

In addition, with Auto mode selected, the result will be expressed in decimal notation if we include a decimal point with one of the integers in the expressions. If we enter $\frac{2}{15} + \frac{7}{12}$ in Auto mode, for example, the result appears in decimal notation.

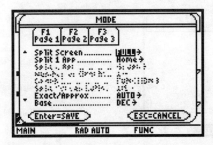
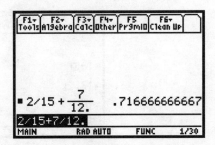

To convert a fractional result to decimal form, use $\boxed{\triangle}$ to move into the history area and highlight the result. Press $\boxed{\text{ENTER}}$ to copy the result to the entry line. Then press $\boxed{\diamond}$ $\boxed{\text{ENTER}}$ to see the decimal form of the result. This occurs regardless of the Exact/Approx setting. (Note that we ordinarily set the calculator in Auto mode.)

SQUARE ROOTS

Section 1.4, Example 5 Graph the real number $\sqrt{3}$ on a number line.

We can use the calculator to find a decimal approximation for $\sqrt{3}$. If the calculator is in Exact or Auto mode, press $\boxed{\text{2nd}}$ $\boxed{\sqrt{}}$ 3 $\boxed{)}$ $\boxed{\diamond}$ $\boxed{\text{ENTER}}$. If the calculator is in Approximate mode, the $\boxed{\diamond}$ is not necessary. Note that the calculator supplies a left parenthesis along with the radical symbol. We must supply the right parenthesis to complete the expression.

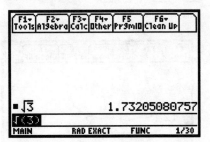

This approximation can now be used to locate $\sqrt{3}$ on the number line. The graph appears on page 33 of the text.

NEGATIVE NUMBERS AND ABSOLUTE VALUE

On the entry line of the TI-89, $|x|$ is written abs(x). Since we selected Pretty Print mode earlier, the traditional notation will appear in the history area.

Section 1.4, Example 8 Find each absolute value: (a) $|-3|$; (b) $|7.2|$; (c) $|0|$.

Absolute value notation is the first item in the graphing calculator's catalog, an alphabetic list of all the commands on the TI-89. When entering $|-3|$, keep in mind that the gray $\boxed{(-)}$ key in the bottom row of the keypad must be used to enter a negative number on the calculator whereas the black $\boxed{-}$ key in the right-hand column of the keypad is used to enter subtraction. Go to the entry line on the home screen and press $\boxed{\text{CATALOG}}$ $\boxed{\text{ENTER}}$ $\boxed{(-)}$ 3 $\boxed{)}$ $\boxed{\text{ENTER}}$. To find $|7.2|$ use the keystrokes above, replacing $\boxed{(-)}$ 3 with 7 $\boxed{.}$ 2. Note that, in order to get the result in decimal form, $\boxed{\diamond}$ must also be pressed before $\boxed{\text{ENTER}}$ if the calculator is set in Exact mode. To find $|0|$ replace $\boxed{(-)}$ 3 with 0. Note that the calculator supplies a left parenthesis along with the absolute value notation. We must supply the right parenthesis to close the absolute-value expression.

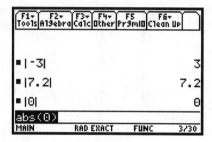

ORDER OF OPERATIONS; EXPONENTS AND GROUPING SYMBOLS

The TI-89 follows the rules for order of operations.

Section 1.8, Example 8 Calculate: $\dfrac{12(9-7)+4\cdot 5}{2^4+3^2}$.

The fraction bar must be replaced with a set of parentheses around the entire numerator and another set of parentheses around the entire denominator when this expression is entered in the calculator. To enter an exponential expression, first enter the base, then use the $\boxed{\wedge}$ key followed by the exponent.

To enter the expression above and express the result in fraction notation, set the calculator in Exact or Auto mode and press $\boxed{(}$ 1 2 $\boxed{(}$ 9 $\boxed{-}$ 7 $\boxed{)}$ $\boxed{+}$ 4 $\boxed{\times}$ 5 $\boxed{)}$ $\boxed{\div}$ $\boxed{(}$ 2 $\boxed{\wedge}$ 4 $\boxed{+}$ 3 $\boxed{\wedge}$ 2 $\boxed{)}$. Remember to use the black $\boxed{-}$ key for subtraction rather than the gray $\boxed{(-)}$ key, which is used to enter negative numbers.

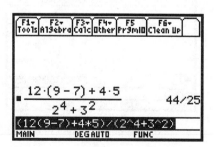

Note that although the $\wedge$ symbol appears in the expression on the entry line, the history area shows the exponent in the traditional format. This happens because we selected Pretty Print mode earlier.

Chapter 2
Equations, Inequalities, and Problem Solving

EDITING ENTRIES

In **Section 2.1** of the text the procedure for editing entries is discussed on page 75. This procedure is described on page 121 of this manual.

EVALUATING FORMULAS; THE TABLE FEATURE: ASK MODE

Section 2.3, Example 2 Use the formula $B = 30a$, described in Example 1, to determine the minimum furnace output for well-insulated houses containing 800 ft^2, 1500 ft^2, 2400 ft^2, and 3600 ft^2.

First we replace B with y and a with x and enter the formula $y = 30x$ on the equation-editor screen as equation y_1. Press $\boxed{\diamond}$ $\boxed{Y=}$ to access this screen. If a plot is currently turned on, it should be turned off, or deselected, now. A check mark beside a plot indicates that it is currently selected. To deselect it, move the cursor to the plot on the equation-editor screen. Then press $\boxed{F4}$. There should now be no check mark beside the plot, indicating that it has been deselected. Do this for each plot that has a check mark beside it. If there is currently an expression displayed for y_1, clear it by positioning the cursor beside "y1 =" and pressing $\boxed{\text{CLEAR}}$. Do the same for expressions that appear on all other "y =" lines by using $\boxed{\triangledown}$ to move to a line and then pressing $\boxed{\text{CLEAR}}$. Then use $\boxed{\triangle}$ or $\boxed{\triangledown}$ to move the cursor beside "y1 =." Now enter $y_1 = 30x$ on the entry line of the equation-editor screen and paste it beside "y1 =" by pressing 3 0 $\boxed{X}$ $\boxed{\text{ENTER}}$.

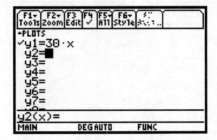

To edit an entry on the equation-editor screen, use $\boxed{\triangle}$ or $\boxed{\triangledown}$ to highlight it and then press $\boxed{\text{ENTER}}$. This copies the entry to the entry line where it can be edited as described on page 121 of this manual.

For an equation entered in the equation-editor screen, a table of x- and y-values can be displayed. We will use a table to evaluate the formula for the given values. Once the equation is entered, press $\boxed{\diamond}$ $\boxed{\text{TblSet}}$ to display the Table Setup screen. (TblSet is the green $\diamond$ operation associated with the $\boxed{F4}$ key.) The Table Setup screen can also be accessed by pressing $\boxed{\diamond}$ $\boxed{\text{TABLE}}$ $\boxed{F2}$. (TABLE is the green $\diamond$ operation associated with the $\boxed{F5}$ key.) You can choose to supply the x-values yourself or you can set the calculator to supply them. To choose the x-values yourself, move the cursor to the "Independent" line. Then press $\boxed{\triangleright}$ 2

ENTER to select Ask mode. In Ask mode the calculator disregards the other settings on the Table Setup screen.

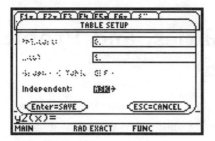

Now press ◇ TABLE to view the table. If you select Ask before a table is displayed for the first time on your calculator, a blank table is displayed. If a table has previously been displayed, the table you now see will continue to show the values in the previous table.

Values for x can be entered in the x-column of the table and the corresponding values for y_1 will be displayed in the y_1-column. To enter 800, 1500, 2400, and 3600 press 8 0 0 ENTER ▽ 1 5 0 0 ENTER ▽ 2 4 0 0 ENTER ▽ 3 6 0 0 ENTER. Any additional x-values that are displayed are from a table that was previously displayed on the Auto setting. The y-values are displayed in the table in scientific notation rounded to one decimal place. (See page 395 in the text for a discussion of scientific notation.) To see the exact values move the cursor to the entries and read the values displayed at the bottom of the screen.

We see that $y_1 = 24,000$ when $x = 800$, $y_1 = 45,000$ when $x = 1500$, $y_1 = 72,000$ when $x = 2400$, and $y_1 = 108,000$ when $x = 3600$, so the furnace outputs for 800 ft^2, 1500 ft^2, 2400 ft^2, and 3600 ft^2 are 24,000 Btu's, 45,000 Btu's, 72,000 Btu's, and 108,000 Btu's, respectively.

THE TABLE FEATURE: AUTO MODE

Section 2.4, Example 8 Village Stationers wants the customer service team to be able to look up the cost c of merchandise being returned when only the total amount paid T (including tax) is shown on the receipts. Use the formula $c = T/1.05$, developed in Example 7, to create a table of values showing cost given a total amount paid. Assume that the least possible sale is $0.21.

We will use the graphing calculator to create the table of values. To enter the formula, we first replace c with y and T with x. Then we enter the formula on the equation-editor screen as $y = x/1.05$. Be sure that the Plots are turned off and that any previous entries are cleared. (See page 125 of this manual for the procedure for turning off the Plots and clearing equations.)

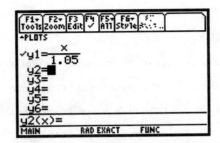

We will create a table in which the calculator supplies the x-values beginning with a value we specify and continuing by adding a value we specify to the preceding value for x. We will begin with an x-value of 0.21 (corresponding to \$0.21) and choose successive increases of 0.01 (corresponding to \$0.01).

To do this, first press $\boxed{\diamond}$ $\boxed{\text{TblSet}}$ or $\boxed{\diamond}$ $\boxed{\text{TABLE}}$ $\boxed{\text{F2}}$ to access the TABLE SETUP window. If "Independent" is set to "Auto" on the Table Setup screen, the calculator will supply values for x, beginning with the value specified as tblStart and continuing by adding the value of Δtbl to the preceding value for x. If the table was previously set to Ask, the blinking cursor will be positioned over ASK. Change this setting to AUTO by pressing $\boxed{\triangleright}$ 1 $\boxed{\text{ENTER}}$. Now the Table Setup screen must once again be accessed so that we can set tblStart and Δtbl. Enter a minimum x-value of 0.21, an increment of 0.01, and a Graph $<->$ Table setting of OFF by first positioning the cursor beside tblStart and then pressing $\boxed{.}$ 2 1 $\boxed{\triangledown}$ $\boxed{.}$ 0 1 $\boxed{\triangledown}$ $\boxed{\triangleright}$ 1 $\boxed{\text{ENTER}}$. Press $\boxed{\diamond}$ $\boxed{\text{TABLE}}$ to see the table.

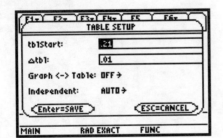

We can change the number of decimal places that the calculator will display to 2 so that the costs are rounded to the nearest cent. To do this, press $\boxed{\text{MODE}}$ to access the MODE screen. Then press $\boxed{\triangledown}$ $\boxed{\triangledown}$ to move the cursor to the third line, Display Digits, and press the $\boxed{\triangleright}$ key to display the options. Then use $\boxed{\triangledown}$ or $\boxed{\triangle}$ to highlight FLOAT 2. Finally, press $\boxed{\text{ENTER}}$ $\boxed{\text{ENTER}}$ to select and save two decimal places on the MODE screen.

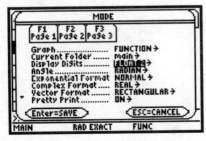
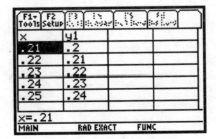

Use the $\boxed{\triangledown}$ and $\boxed{\triangle}$ keys to scroll through the table.

Before proceeding, return to the MODE screen and reselect "Float." This allows the number of decimal places to vary, or float, according to the computation being performed.

In **Section 2.5, Example 2**, two equations are entered on the equation-editor screen. The first equation, $y_1 = x + 1$, can be entered as described on page 125 of this manual. Entering y_1 moves the cursor beside "y2 =." Now enter the right-hand side of the second equation, $x + (x + 1)$. Note that the equation-editor screen will not display the parentheses even if they are keyed in on the entry line.

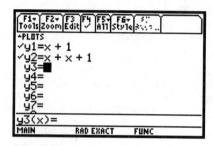

Chapter 3
Introduction to Graphing and Equations

SETTING THE VIEWING WINDOW

The viewing window is the portion of the coordinate plane that appears on the calculator's screen. It is defined by the minimum and maximum values of x and y: xmin, xmax, ymin, and ymax. The notation [xmin, xmax, ymin, ymax] is used in the text to represent these window settings or dimensions. For example, $[-12, 12, -8, 8]$ denotes a window that displays the portion of the x-axis from -12 to 12 and the portion of the y-axis from -8 to 8. In addition, the distance between tick marks on the axes is defined by the settings xscl and yscl. In this manual xscl and yscl will be assumed to be 1 unless noted otherwise. The setting xres sets the pixel resolution. We usually select xres = 2. The window corresponding to the settings $[-20, 30, -12, 20]$, xscl = 5, yscl = 2, xres = 2, is shown below.

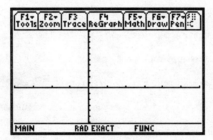

Press ◇ WINDOW to display the current window settings on your calculator. (WINDOW is the ◇ operation associated with the F2 key on the top row of the keypad.) The standard settings are shown below.

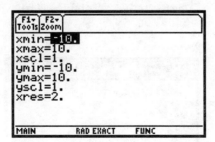

Section 3.1, Example 5 Set up a $[-100, 100, -5, 5]$ viewing window on a graphing calculator, choosing appropriate scales for the axes.

To change a setting, position the cursor beside the setting you wish to change and enter the new value. We will enter the given settings and let xscl = 10 and yscl = 1. The choice of scales for the axes may vary. We select scaling that will allow some space between tick marks. If a scale that is too small is chosen, the tick marks will blend and blur. To change the settings to $[-100, 100, -5, 5]$, Xscl = 10, on the Window screen, start with the setting beside xmin = highlighted and press (−) 1 0 0

ENTER 1 0 0 ENTER 1 0 ENTER (−) 5 ENTER 5 ENTER 1 ENTER. The ▽ key may be used instead of ENTER after typing each window setting. To see the window, press ◊ GRAPH. (GRAPH is the ◊ operation associated with the F3 key on the top row of the keypad.)

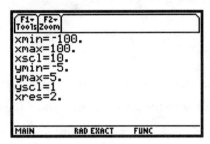 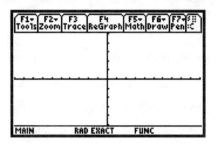

QUICK TIP: To return quickly to the standard window setting $[-10, 10, -10, 10]$, xscl = 1, yscl = 1, when either the Window screen or the Graph screen is displayed, press F2 to access the ZOOM menu and then press 6 to select item 6, ZoomStd (Zoom Standard).

GRAPHING EQUATIONS

After entering an equation and setting a viewing window, you can view the graph of an equation.

Section 3.2, Example 4 Graph $y = 2x$ using a graphing calculator.

Equations are entered on the equation-editor screen. Press ◊ Y = to access this screen. If Plot 1 was used in the example above, it should be turned off, or deselected, now. A check mark beside Plot 1 indicates that it is currently selected. To deselect it, move the cursor to Plot 1 on the equation-editor screen. Then press F4. There should now be no check mark beside Plot 1, indicating that it has been deselected. If there is currently an expression displayed for y_1, clear it as described above. Do the same for expressions that appear on all other "y =" lines by using ▽ to move to a line and then pressing CLEAR. Then use △ or ▽ to move the cursor beside "y1 =." Now enter $y_1 = 2x$ on the entry line of the equation-editor screen and paste it beside "y1 =" by pressing 2 X ENTER.

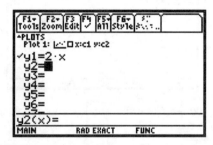

The standard $[-10, 10, -10, 10]$ window is a good choice for this graph. Either enter these dimensions in the WINDOW screen and then press ◊ GRAPH to see the graph or simply press F2 6 to select the standard window and see the graph.

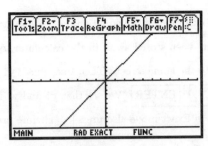

To edit an entry on the equation-editor screen, use △ or ▽ to highlight it and then press ENTER. This copies the entry to the entry line where it can be edited as described on page 121 of this manual.

SOLVING EQUATIONS GRAPHICALLY: THE INTERSECT METHOD

We can use the Intersection feature from the Math menu on the Graph screen to solve equations.

Section 3.3, Example 2 Solve using a graphing calculator: $-\frac{3}{4}x + 6 = 2x - 1$.

On the equation-editor screen, clear any existing entries and then enter $y_1 = -\frac{3}{4}x + 6$ and $y_2 = 2x - 1$. Press F2 6 to graph these equations in the standard viewing window. The solution of the equation $-\frac{3}{4}x + 6 = 2x - 1$ is the first coordinate of the point of intersection of these graphs. To use the Intersection feature to find this point, first press F5 5 to select Intersection from the Math menu on the Graph screen. The query "1st curve?" appears at the bottom of the screen. The blinking cursor is positioned on the graph of y_1. This is indicated by the 1 in the upper right-hand corner of the screen. Press ENTER to indicate that this is the first curve involved in the intersection. Next the query "2nd curve?" appears at the bottom of the screen. The blinking cursor is now positioned on the graph of y_2 and the notation 2 should appear in the top right-hand corner of the screen. Press ENTER to indicate that this is the second curve. We identify the curves for the calculator since we could have more than two graphs on the screen at once. After we identify the second curve, the query "Lower bound?" appears at the bottom of the screen. Use the right and left arrow keys to move the blinking cursor to a point to the left of the point of intersection of the lines or type an x-value less than the x-coordinate of the point of intersection. Then press ENTER. Next the query "Upper bound?" appears. We give a lower and an upper bound since some pairs of curves have more than one point of intersection. Move the cursor to a point to the right of the point of intersection or type an x-value greater than the x-value of the point of intersection and press ENTER. Now the coordinates of the point of intersection appear at the bottom of the screen.

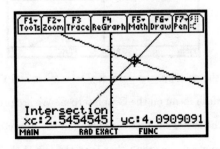

We see that $x = 2.5454545$, so the solution of the equation is 2.5454545.

We can check the solution by evaluating both sides of the equation $-\frac{3}{4}x + 6 = 2x - 1$ for this value of x. The first coordinate of the point of intersection has automatically been stored as xc in the calculator, so we evaluate y1 and y2 for this value of x. First press $\boxed{\text{HOME}}$ or $\boxed{\text{2nd}}$ $\boxed{\text{QUIT}}$ to go to the home screen. Then to evaluate y1 press $\boxed{\text{Y}}$ $\boxed{1}$ $\boxed{(}$ $\boxed{\text{X}}$ $\boxed{\text{alpha}}$ $\boxed{\text{C}}$ $\boxed{)}$ $\boxed{\text{ENTER}}$. To evaluate y_2 press $\boxed{\text{Y}}$ $\boxed{2}$ $\boxed{(}$ $\boxed{\text{X}}$ $\boxed{\text{alpha}}$ $\boxed{\text{C}}$ $\boxed{)}$ $\boxed{\text{ENTER}}$. We see that y1 and y2 have the same value when $x = 2.5454545$, so the solution checks. If your calculator is set in Exact mode also press $\boxed{\diamond}$ before pressing $\boxed{\text{ENTER}}$ in order to see the result in decimal notation.

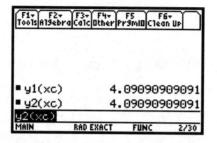

Note that although the procedure above verifies that 2.5454545 is the solution, it is actually an approximation of the solution. To find the exact solution we can solve the equation algebraically.

SOLVING EQUATIONS GRAPHICALLY: THE ZERO METHOD

As an alternative to the Intersect method, we can use the Zero feature from the Math menu on the Graph screen to solve equations.

Section 3.3, Example 7 Solve $3 - 8x = 5 - 7x$ using the zero method.

First we get zero on one side of the equation:
$$3 - 8x = 5 - 7x$$
$$-2 - 8x = -7x$$
$$-2 - x = 0.$$

Then we graph $y = -2 - x$. We will use the standard window.

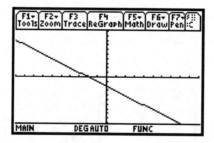

The first coordinate of the x-intercept of the graph is the zero of $y = -2 - x$ and thus is the solution of the original equation. Press $\boxed{\text{2nd}}$ $\boxed{\text{F5}}$ 2 to select Zero from the Math menu on the Graph screen. We are prompted to select a lower bound. This means that we must choose an x-value that is to the left of the first coordinate of the x-intercept. This can be done by using the left and right arrow keys to move the cursor to a point on the graph that is to the left of the intercept or by keying in an x-value that is less than the first coordinate of the intercept.

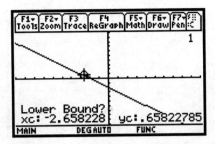

Once this is done, press ENTER. Now we are prompted to select an upper bound that is to the right of the x-intercept. Again we can use the arrow keys or key in a value.

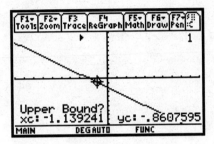

Press ENTER again. We see that $y = 0$ when $x = -2$, so -2 is the zero of $y = -2 - x$ and is thus the solution of the original equation, $3 - 8x = 5 - 7x$.

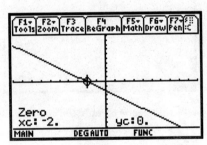

EVALUATING A FUNCTION

Function values can be found in several different ways on the TI-89.

Section 3.4, Example 5 For $f(a) = 2a^2 - 3a + 1$, find $f(3)$ and $f(-5.1)$.

One method for finding function values involves using function notation directly. To do this, first press ◇ Y= and clear any entries that are present. Also be sure the plots are deselected. (See page 125 of this manual for instructions for clearing equations and deselecting the plots.) Now enter the function on the equation-editor screen. Mentally replace a with x and $f(a)$ with y_1. Then enter $y_1 = 2x^2 - 3x + 1$. Now, to find $f(3)$, or y1(3), directly first press HOME or 2nd QUIT to go to the home screen. Then enter "y1(3)" on the entry line by pressing Y 1 (3). Finally press ENTER. We see that $y_1(3) = 10$, or $f(3) = 10$.

To find $f(-5.1)$, or $y_1(-5.1)$, we can repeat the previous procedure using -5.1 in place of 3, or we can edit the previous entry. To edit, press ▷ to go to the right side of the previous entry on the entry line. Then press ◁ ← to delete the 3. Now press (−) 5 . 1 to replace 3 with -5.1. Finally press ENTER to find that $y_1(-5.1) = 68.32$, or $f(-5.1) = 68.32$.

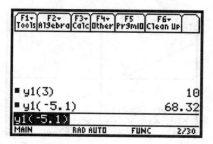

We can also find function values from the graph of the function.

Section 3.4, Example 6 Find $g(2)$ for $g(x) = 2x - 5$.

First press $\boxed{\diamond}$ $\boxed{Y=}$ to go to the equation editor screen and then clear any entries that are present. Also be sure that the plots are turned off. (See page 125 of this manual for instructions for clearing equations and turning off plots.) Now enter $y_1 = 2x - 5$ and press $\boxed{F2}$ 6 to graph this function in the standard viewing window. We will use the Value feature from the Math menu on the Graph screen to find the value of y_1 when $x = 2$. This is $g(2)$. Press $\boxed{F5}$ 1 to select Value. Press 2 $\boxed{\text{ENTER}}$. We now see xc:2, yc:-1 at the bottom of the screen. This indicates that when the x-coordinate on this graph is 2, the corresponding y-coordinate is -1, so $g(2) = -1$.

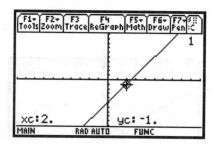

When using the Value feature, note that the x-value entered must be in the viewing window. That is, x must be a number between xmin and xmax.

Chapter 4
Linear Equations, Inequalities, and Graphs

SQUARING THE VIEWING WINDOW

Section 4.4, Example 9 Determine whether the lines given by the equations $3x - y = 7$ and $x + 3y = 1$ are perpendicular, and check by graphing.

In the text each equation is solved for y in order to determine the slopes of the lines. We have $y = 3x - 7$ and $y = -\frac{1}{3}x + \frac{1}{3}$. Since $3\left(-\frac{1}{3}\right) = -1$, we know that the lines are perpendicular. To check this, we graph y1 $= 3x - 7$ and y2 $= -\frac{1}{3}x + \frac{1}{3}$. The graphs are shown on the right below in the standard viewing window.

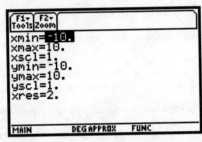 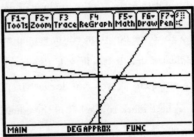

Note that the graphs do not appear to be perpendicular. This is due to the fact that, in the standard window, the distance between tick marks on the y-axis is about 1/2 the distance between tick marks on the x-axis. It is often desirable to choose window dimensions for which these distances are the same, creating a "square" window. On the TI-89 any window in which the ratio of the length of the y-axis to the length of the x-axis is 1/2 will produce this effect. This can be accomplished by selecting dimensions for which ymax $-$ ymin $= \frac{1}{2}$(xmax $-$ xmin). For example, the windows $[-12, 12, -6, 6]$ and $[-6, 6, -3, 3]$ are square.

When we change the window dimensions to $[-12, 12, -6, 6]$ and press ◇ GRAPH , the graphs now appear to be perpendicular as shown on the right below. From the equation-editor, Window, or Graph screen, we could also press F2 5 to select ZoomSqr. When this is done, the calculator will select a square window.

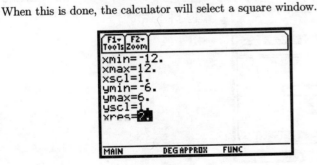 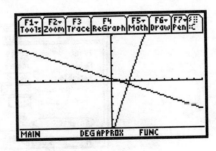

ENTERING AND PLOTTING DATA; LINEAR REGRESSION

We can use the Linear Regression feature on the TI-89 to fit a linear equation to a set of data.

Section 4.5, Example 9 The amount of paper recovered in the United States for various years is shown in the following table.

Years	Amount of Paper Recovered (in millions of tons)
1988	26.2
1990	29.1
1992	34.0
1994	39.7
1996	43.1
1998	45.1
2000	49.4

(a) Fit a linear function to the data.

(b) Graph the function and use it to estimate the amount of paper that will be recovered in 2003.

(a) First press ◇ Y = to go to the equation-editor screen and clear any equations that are currently selected. A check mark to the left of an equation indicates that it is selected. (See page 125 of this manual for the procedure for clearing equations.) If you wish, instead of clearing an equation, you can deselect it. To do this, position the cursor beside the = sign and press F4. Note that there is no longer a check mark to the left of the equation, indicating that the equation has been deselected. The graph of an equation that has been deselected will not appear when ◇ GRAPH is pressed. A deselected equation can be selected again by positioning the cursor beside the = sign and pressing F4. Note that a check mark once again appears to the left of the equation.

We will enter the coordinates of the ordered pairs in the Data/Matrix editor. Press APPS 6 3 to display the new data variable screen in the Data/Matrix editor. We must now enter a data variable name in the Variable box on this screen. The name can contain from 1 to 8 characters and cannot start with a numeral. Some names are preassigned to other uses on the TI-89. If you try to use one of these, you will get an error message. Press ▽ ▽ to move the cursor to the Variable box. We will name our data variable "news." To enter this name, first lock the alphabetic keys on by pressing 2nd a-lock. Then press N E W S. Note that N, E, W, and S are the purple alphabetic operations associated with the 6, ÷, ·, and 3 keys, respectively.

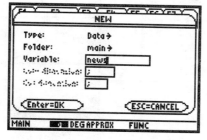

After typing the name of the data variable, unlock the alphabetic keys by pressing the purple alpha key. Now press ENTER ENTER to go to the data-entry screen. Assuming the data variable name "news" has not previously been used in your calculator, this screen will contain empty data lists with row 1, column 1 highlighted. If entries have previously been made in a data variable named "news," they can be cleared by pressing $\boxed{F1}$ 8 $\boxed{ENTER}$.

We will enter the first coordinates (x-coordinates) of the points, as the number of years since 1988, in column c1 and the second coordinates (y-coordinates), in millions of tons, in c2. To enter the first x-coordinate, 0 (for 1988), press 0 $\boxed{ENTER}$. Continue typing the x-values 2, 4, 6, 8, 10, and 12, each followed by $\boxed{ENTER}$. The entries can be followed by $\boxed{\triangledown}$ rather than $\boxed{ENTER}$ if desired. Press $\boxed{\triangleright}$ $\boxed{\triangle}$ $\boxed{\triangle}$ $\boxed{\triangle}$ $\boxed{\triangle}$ to move to the top of column c2. Type the y-values 26.2, 29.1, 34.0, 39.7, 43.1, 45.1, and 49.4 in succession, each followed by $\boxed{ENTER}$ or $\boxed{\triangledown}$. Note that the coordinates of each point must be in the same position in both lists.

Now use the calculator's linear regression feature to fit a linear equation to the data. Press $\boxed{F5}$ to display the Calculate menu. Press $\boxed{\triangleright}$ 5 to select LinReg (linear regression). Then press $\boxed{\triangledown}$ $\boxed{alpha}$ $\boxed{C}$ 1 $\boxed{\triangledown}$ $\boxed{alpha}$ $\boxed{C}$ 2 to indicate that the data in c1 and c2 will be used for x and y, respectively. Press $\boxed{\triangledown}$ $\boxed{\triangleright}$ $\boxed{\triangledown}$ $\boxed{ENTER}$ to indicate that the regression equation should be copied to the equation-editor screen as y1. Finally press $\boxed{ENTER}$ again to see the STAT VARS screen which displays the coefficients a and b of the regression equation $y = ax + b$. We see that the regression equation is $y = 1.976786x + 26.225$.

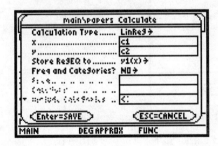
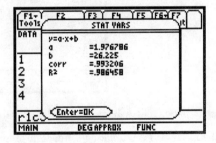

Note that values for "corr" (the correlation coefficient) and r^2 (the coefficient of determination) will also be displayed. These numbers indicate how well the regression line fits the data. While it is possible to suppress these numbers on some graphing calculators, this cannot be done on the TI-89.

(b) Now we will graph the regression equation. In order to see the data points along with the graph of the equation we will turn on and define a plot. To do this, from the STAT VARS screen first press $\boxed{ENTER}$ $\boxed{F2}$ to go to the Plot Setup screen. We will use Plot 1, which is highlighted. If any plot settings are currently entered beside "Plot 1," clear them by pressing $\boxed{F3}$. Clear settings shown beside any other plots as well by using $\boxed{\triangledown}$ to highlight each plot in turn and then pressing $\boxed{F3}$.

Now we define Plot 1. Use $\boxed{\triangle}$ to highlight Plot 1 if necessary. Then press $\boxed{F1}$ to display the Plot Definition screen. The item on the first line, Plot Type, is highlighted. We will choose a scatter diagram, denoted by Scatter, by pressing ▷ 1. Now press ▽ to go to the next line, Mark. Here we select the type of mark or symbol that will be used to plot the points. We select a box by pressing ▷ 1. Now we must tell the calculator which columns of the data variable to use for the x- and y-coordinates of the points to be plotted. Press ▽ to move the cursor to the "x" line and enter c1 as the source of the x-coordinates by pressing alpha C 1. (C is the purple alphabetic operation associated with the) key.) Press ▽ alpha C 2 to go the the "y" line and enter c2 as the source of the y-coordinates.

To select the dimensions of the viewing window notice that the years in the table range from 0 to 12 and the number of millions of tons of paper ranges from 26.2 to 49.4. We want to select dimensions that will include all of these values. One good choice is [0, 15, 0, 60], yscl = 10. Enter these dimensions in the Window screen. Then press ◇ GRAPH to see the graph of the regression line on the same axes as the data.

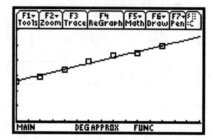

You can press ◇ Y =, if desired, to see the regression equation entered as y1 on the equation-editor screen.

To estimate the amount of paper that will be recovered in 2003, evaluate the regression equation for $x = 15$. (2003 is 15 years after 1988.) Use any of the methods for evaluating a function presented earlier in this chapter. (See pages 133 and 134.) We will use the Value feature from the Math menu on the Graph screen.

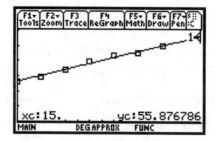

When $x = 15, y \approx 55.9$, so we estimate that there will be about 55.9 million tons of paper recovered in 2003.

TI-89

THE TRACE FEATURE

Section 4.6, Example 10 Find the domain of $f+g$ if $f(x)=\sqrt{2x-5}$ and $g(x)=\sqrt{x+1}$. In the text it is found that the domain of $f+g$ is $\left\{x \middle| x \geq \frac{5}{2}\right\}$, or $\left[\frac{5}{2}, \infty\right)$. This can be confirmed, at least approximately, by tracing the graph of $f+g$. First graph $y=\sqrt{2x-5}+\sqrt{x+1}$. We will use the standard window. Now press $\boxed{F3}$ to select Trace and use the right and left arrow keys to move the cursor along the curve. We see that no y-values are given for x-values less than 2.5, or $\frac{5}{2}$. This indicates that x-values less than 2.5 are not in the domain of $f+g$. As we move to the right, ordered pairs appear to extend without bound confirming that the domain of $f+g$ is $\left\{x \middle| x \geq \frac{5}{2}\right\}$, or $\left[\frac{5}{2}, \infty\right)$.

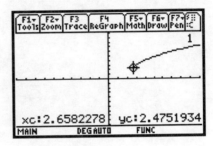

Chapter 5
Polynomials

CHECKING OPERATIONS ON POLYNOMIALS

A graphing calculator can be used to check operations on polynomials.

Section 5.3, Interactive Discovery Check the addition $(-3x^3 + 2x - 4) + (4x^3 + 3x^2 + 2) = x^3 + 3x^2 + 2x - 2$.

There are several ways in which we can use a graphing calculator to check this result. One of these is to compare the graphs of $y1 = (-3x^3 + 2x - 4) + (4x^3 + 3x^2 + 2)$ and $y2 = x^3 + 3x^2 + 2x - 2$. This is most easily done when different graph styles are used for the graphs.

Eight graph styles can be selected on the TI-89. The **path graph style** can be used, along with the line style, to determine whether graphs coincide. First, on the equation-editor screen, enter $y1 = (-3x^3+2x-4)+(4x^3+3x^2+2)$ and $y2 = x^3+3x^2+2x-2$. We will select the path style from the style menu for $y2$. To do this, highlight the expression for $y2$ and then press $\boxed{\text{2nd}}$ $\boxed{\text{F6}}$ 6 $\boxed{\text{ENTER}}$ or press $\boxed{\text{2nd}}$ $\boxed{\text{F6}}$ $\boxed{\triangledown}$ $\boxed{\triangledown}$ $\boxed{\triangledown}$ $\boxed{\triangledown}$ $\boxed{\triangledown}$ $\boxed{\text{ENTER}}$.

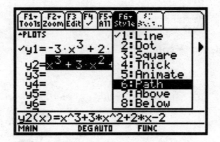

The calculator will graph $y1$ first as a solid line. Then $y2$ will be graphed as the circular cursor traces the leading edge of the graph, allowing us to determine visually whether the graphs coincide. In this case, the graphs appear to coincide, so the factorization is probably correct.

We can also check the addition by **subtracting** the result from the original sum. With $y1$ and $y2$ entered as described above, position the cursor beside "$y3 =$" and enter $y3 = y1 - y2$ by pressing $\boxed{\text{Y}}$ 1 $\boxed{-}$ $\boxed{\text{Y}}$ 2 $\boxed{\text{ENTER}}$. If the subtraction is correct, $y1 = y2$, so $y3 = 0$ and the graph of $y3$ will be $y = 0$, or the x-axis. Since we are interested only in values of $y3$, deselect $y1$ and $y2$ as described on page 136 of this manual. Select the path graph style for $y3$ as described above.

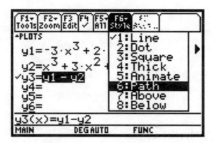

Now press ◇ GRAPH and determine if the graph of $y3$ is traced over the x-axis. Since it is, the sum is correct.

We can also use a table of values to **compare values** of $y1$ and $y2$. If the expressions for $y1$ and $y2$ are the same for each given x-value, the result checks. If you deselected $y1$ and $y2$ to check the sum using subtraction as described above, select them again now. Then look at a table set in Auto mode. (See page 126.) Since the values of $y1$ and $y2$ are the same for each given x-value, the result checks. Scrolling through the table to look at additional values makes this conclusion more certain.

We can also check the sum using a **horizontal split screen**, or **top-bottom split screen**. We will choose to display the graph in the top half of the screen and a table of values in the bottom half.

First enter $y1$ and $y2$ as described above. Then set up a table in Auto mode. We will use tblStart $= -3$ and Δtbl $= 1$. To select the top-bottom split-screen option, first press MODE F2 to access the second page of the Mode screen. Then press ▷ 2 to select a screen split into a top and a bottom portion. Next press ▽ ▷ 4 ▽ ▷ 5 ENTER to select the graph to be displayed in the top portion of the screen and a table of values in the bottom portion. It might be necessary to press ◇ TABLE to see the table. We show the functions graphed in the standard window.

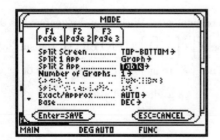 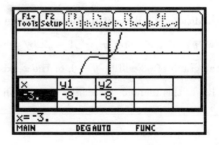

We can also use a **vertical split screen**, of **left-right split screen** to check the addition. First enter $y1$, $y2$, and $y3$ as described above and deselect $y1$ and $y2$. Then press MODE F2 to display the second Mode screen. Press ▷ to select left-right. Then select Graph for Split 1 App and Table for Split 2 App. Now press ENTER to see the graph of $y3$ on the left side of the screen and a table of values for $y3$ on the right side. As we did above, we use show the graph in the standard window and use a table set in Auto mode with tblStart $= -3$ and Δtbl $= 1$. Since the graph appears to be $y = 0$, or the x-axis, and all of the $y3$-values in the table are 0, we confirm that the result is correct.

In order to return to a full screen, return to the second Mode screen and select Full for the Split Screen option.

EVALUATING POLYNOMIALS IN SEVERAL VARIABLES

Section 5.6, Example 1 Evaluate the polynomial $4 + 3x + xy^2 + 8x^3y^3$ for $x = -2$ and $y = 5$.

To evaluate a polynomial in two or more variables, substitute numbers for the variables. This can be done by substituting values for x and y directly or by storing the values of the variables in the calculator. The procedure for direct substitution is described on page 120 of this manual.

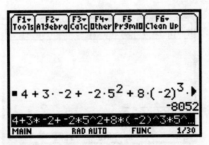

To evaluate the polynomial by first storing -2 as x and 5 as y, we proceed as follows. To store -2 as x, press $\boxed{(-)}$ 2 $\boxed{\text{STO}\triangleright}$ $\boxed{X}$ $\boxed{\text{ENTER}}$. Now store 5 as y. Press 5 $\boxed{\text{STO}\triangleright}$ $\boxed{Y}$ $\boxed{\text{ENTER}}$. Next enter the algebraic expression Press 4 $\boxed{+}$ 3 $\boxed{\times}$ $\boxed{X}$ $\boxed{+}$ $\boxed{X}$ $\boxed{\times}$ $\boxed{Y}$ $\boxed{\wedge}$ 2 $\boxed{+}$ 8 $\boxed{\times}$ $\boxed{X}$ $\boxed{\wedge}$ 3 $\boxed{\times}$ $\boxed{Y}$ $\boxed{\wedge}$ 3. Finally, press $\boxed{\text{ENTER}}$ to find the value of the expression.

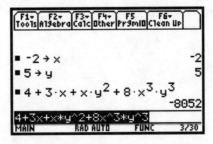

SCIENTIFIC NOTATION

To enter a number in scientific notation, first type the decimal portion of the number; then press the $\boxed{\text{EE}}$ key in the left column of the keypad; finally type the exponent, which can be at most three digits. For example, to enter 1.789×10^{-11} in scientific notation, press $1\ \boxed{.}\ 7\ 8\ 9\ \boxed{\text{EE}}\ \boxed{(-)}\ 1\ 1\ \boxed{\text{ENTER}}$. To enter 6.084×10^{23} in scientific notation, press $6\ \boxed{.}\ 0\ 8\ 4\ \boxed{\text{EE}}\ 2\ 3\ \boxed{\text{ENTER}}$. The decimal portion of each number appears before a small E while the exponent follows the E.

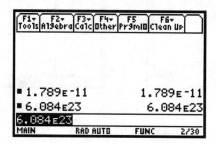

The graphing calculator can be used to perform computations in scientific notation.

Section 5.8, Example 9 Use a graphing calculator to check the computation $(7.2 \times 10^{-7}) \div (8.0 \times 10^6) = 9.0 \times 10^{-14}$.

We enter the computation in scientific notation on the entry line of the home screen. Press $7\ \boxed{.}\ 2\ \boxed{\text{EE}}\ \boxed{(-)}\ 7\ \boxed{\div}\ 8\ \boxed{\text{EE}}\ 6\ \boxed{\text{ENTER}}$. We have 9×10^{-14}, which checks.

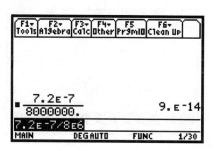

Chapter 6
Polynomials and Factoring

THE INTERSECT AND ZERO METHODS

Section 6.1, Example 1 Solve: $x^2 = 6x$.

To solve this equation by finding the x-coordinates of the points of intersection of $y_1 = x^2$ and $y_2 = 6x$, see the intersect method described on page 131 of this manual.

To solve this equation by writing it as $x^2 - 6x = 0$, graphing $f(x) = x^2 - 6x$, and then finding the values of x for which $f(x) = 0$, or the first coordinates of the x-intercepts of the graph, see the zero method described on page 132 of this manual.

POLYNOMIAL REGRESSION

The TI-89 has the capability to use regression to fit nonlinear polynomial equations to data.

Section 6.78, Example 4(a) The number of bachelor's degrees earned in the biological and life sciences for various years is shown in the following table. Fit a polynomial function of degree 3 (cubic) to the data.

Year	Number of Bachelor's Degrees Earned in Biological/Life Science
1971	35,743
1976	54,275
1980	46,370
1986	38,524
1990	37,204
1994	51,383
2000	63,532

First enter the data in the Data/Matrix editor with the number of years since 1970 in c1 and the number of degree earned, in thousands, in c2. (See page 136 of this manual.) A polynomial in one variable of degree 3 is called a cubic, so we will select cubic regression, denoted CubicReg, from the Calculate menu. This is item 3 under CalculationType. The calculator returns the coefficients for a cubic function of the form $f(x) = ax^3 + bx^2 + cx + d$. Rounding the coefficients to the nearest thousandth, we have $f(x) = 0.009x^3 - 0.371x^2 + 4.176x + 34.415$. We can see the coefficients given to more decimal places on the equation-editor screen, if desired.

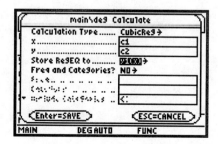 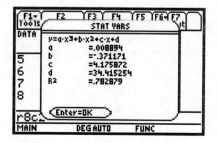

We can use methods discussed earlier in this manual to estimate and predict function values and to find the year or years in which a specific function value occurs.

Chapter 7
Rational Equations and Functions

GRAPHING IN DOT MODE

Consider the graph of the function $T(t) = \dfrac{t^2 + 5t}{2t + 5}$ in Section 7.1, Example 1. Enter $y = (x^2 + 5x)/(2x + 5)$ and graph it in the window $[-5, 5, -5, 5]$.

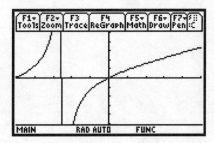

Note that a vertical line that is not part of the graph appears on the screen along with the two branches of the graph. The reason for this is discussed in the text.

This line will not appear if we change from Line style to Dot style. After entering $y1 = (x^2 + 5x)/(2x + 5)$, highlight $y1$ on the equation-editor screen. Then access the Style menu by pressing $\boxed{\text{2nd}}$ $\boxed{\text{F6}}$. Press 2 to select Dot style. Now press $\boxed{\diamond}$ $\boxed{\text{GRAPH}}$ to see the graph of the function in Dot style.

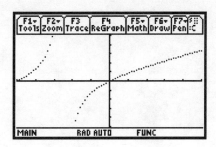

Chapter 8
Systems of Linear Equations and Problem Solving

SOLVING SYSTEMS OF EQUATIONS GRAPHICALLY

We can use the Intersection feature from the Math menu on the Graph screen of the TI-89 to solve a system of two equations in two variables.

Section 8.1, Example 4(a) Solve graphically:

$y - x = 1,$

$y + x = 3.$

We graph the equations in the same viewing window and then find the coordinates of the point of intersection. Remember that equations must be entered in "$y =$" form on the equation-editor screen, so we solve both equations for y. We have $y = x + 1$ and $y = -x + 3$. Enter these equations, graph them in the standard viewing window, and find their point of intersection as described on page 131 of this manual. We see that the solution of the system of equations is $(1, 2)$.

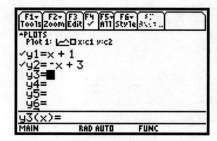 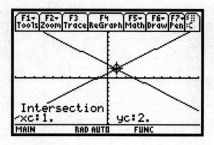

MODELS

Sometimes we model two situations with linear functions and then want to find the point of intersection of their graphs.

Section 8.1, Example 6 (d), (e) The numbers of U. S. travelers to Canada and to Europe are listed in the following table.

Year	U. S. Travelers to Canada (in millions)	U. S. Travelers to Europe (in millions)
1992	11.8	7.1
1994	12.5	8.2
1996	12.9	8.7
1998	14.9	11.1
2000	15.1	13.4

(d) Use linear regression to find two linear equations that can be used to estimate the number of U. S. travelers to Canada and Europe, in millions, x years after 1990.

(e) Use the equations found in part (d) to estimate the year in which the number of U. S. travelers to Europe will be the same as the number of U. S. travelers to Canada.

(d) Enter the data in the Data/Matrix editor as described on page 136 of this manual. We will express the years as the number of years since 1990 (in other words, 1990 is year 0) and enter them in c1. Then enter the number of U. S. travelers to Canada, in millions, in c2 and the number of U. S. travelers to Europe, in millions, in c3.

Now use linear regression to fit a linear function to the data in c1 and c2. The function should also be copied to the equation-editor screen. We will copy it as y1. See page 137 of this manual for the procedure to follow. We get $y1 = 0.45x + 10.74$.

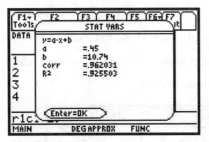
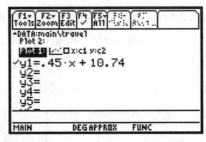

Next we fit a linear function to the data in c1 and c3, entering c1 as x and c3 as y on the Calculate screen. This function will be copied to the equation-editor screen as y2. To enter c3 as y, position the blinking cursor in the y box and press [alpha] [C] 3. To select y2 as the function to which the regression equation will be copied, position the cursor beside "Store Reg EQ to," press [▷], use the [▽] key to highlight y2(x) and press [ENTER]. We get $y2 = 0.775x + 5.05$.

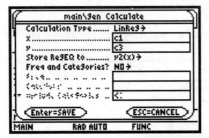
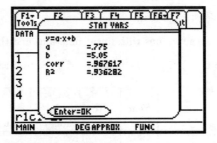

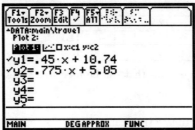

(e) To estimate the year in which the number of U. S. travelers to Europe will be the same as the number of U. S. travelers to Canada, we solve the system of equations

$$y = 0.45x + 10.74,$$
$$y = 0.775x + 5.05.$$

We graph the equations in the same viewing window and then use the Intersection feature to find their point of intersection. Through a trial-and-error process we find that $[0, 30, 0, 30]$, xscl $= 2$, yscl $= 2$, provides a good window in which to see this point. We see that the solution of the system of equations is approximately $(17.51, 18.62)$, so the number of U. S. travelers to Europe will be the same as the number of U. S. travelers to Canada about 17.5 years after 1990, or in 2008.

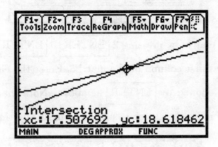

ELIMINATION USING MATRICES

Matrices with up to 999 rows and 99 columns can be entered on the TI-89. The row-equivalent operations necessary to write a matrix in row-echelon form or reduced row-echelon form can be performed on the calculator, or we can go directly to row-echelon form or reduced row-echelon form with a single command. We will illustrate the direct approach for finding reduced row-echelon form.

Section 8.6, Example 1 Solve the following system using a graphing calculator:

$$2x + 5y - 8z = 7,$$
$$3x + 4y - 3z = 8,$$
$$5y - 2x = 9.$$

First we rewrite the third equation in the form $ax + by + cz = d$:

$$2x + 5y - 8z = 7,$$
$$3x + 4y - 3z = 8,$$
$$-2x + 5y = 9.$$

Then we enter the coefficient matrix

$$\begin{bmatrix} 2 & 5 & -8 & 7 \\ 3 & 4 & -3 & 8 \\ -2 & 5 & 0 & 9 \end{bmatrix}$$

in the Data/Matrix editor. We will call the Matrix **a**. Press $\boxed{\text{APPS}}$ 6 3 $\boxed{\triangleright}$ 2 $\boxed{\triangledown}$ $\boxed{\triangledown}$ $\boxed{\text{alpha}}$ $\boxed{A}$ $\boxed{\triangledown}$ 3 $\boxed{\triangledown}$ 4 $\boxed{\text{ENTER}}$ $\boxed{\text{ENTER}}$ to go to the Data/Matrix editor and set up a matrix named A with 3 rows and 4 columns. If a matrix named **a** has previously been saved in your calculator, an error message will be displayed. If this happens, you can select a different name for the matrix we are about to enter or you can delete the current matrix **a** and then enter the new matrix as **a**. To delete a matrix press 2nd VAR-LINK , use ▽ to highlight the name of the matrix being deleted, and then press F1 1 ENTER . (VAR-LINK is the second operation associated with the − key.)

Enter the elements of the first row of the matrix by pressing 2 ENTER 5 ENTER (−) 8 ENTER 7 ENTER . The cursor moves to the element in the second row and first column of the matrix. Enter the elements of the second and third rows of the augmented matrix by typing each in turn followed by ENTER as above. Note that the screen only displays three columns of the matrix. The arrow keys can be used to move the cursor to any element at any time.

Matrix operations are performed on the home screen and are found on the Math Matrix menu. Press HOME or 2nd QUIT to leave the matrix editor and go to this screen. Access the Math Matrix menu by pressing 2nd MATH 4. (MATH is the second operation associated with the 5 numeric key.) The reduced row-echelon form command is item 4 on this menu. Copy it to the entry line of the home screen by pressing 4. We see the command "rref(" on the entry line.

Since we want to find reduced row-echelon form for matrix **a**, we enter **a** by pressing alpha A) . Finally press ENTER to see reduced row echelon form of the original matrix. We see that the solution of the system of equations is $\left(\frac{1}{2}, 2, \frac{1}{2}\right)$ (or (0.5, 2, 0.5) if Auto or Approximate mode is selected rather than Exact mode).

TI-89

EVALUATING DETERMINANTS

We can evaluate determinants using the "det" operation from the MATRIX MATH menu.

Section 8.7, Example 3 Evaluate: $\begin{vmatrix} -1 & 0 & 1 \\ -5 & 1 & -1 \\ 4 & 8 & 1 \end{vmatrix}$.

First enter the 3 x 3 matrix

$$\begin{bmatrix} -1 & 0 & 1 \\ -5 & 1 & -1 \\ 4 & 8 & 1 \end{bmatrix}$$

as described on pages 151 and 152 of this manual. We will enter it as matrix **a**.

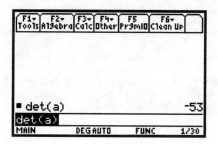

Then press HOME or 2nd QUIT to go to the home screen. Next press 2nd MATH 4 to access the Math Matrix menu. Press 2 to copy the "det(" operation to the entry line of the home screen. Then enter the matrix name **a** by pressing alpha A). Finally, press ENTER to find the value of the determinant of matrix **a**.

INEQUALITIES IN TWO VARIABLES

The solution set of an inequality in two variables can be graphed on the TI-89.

Section 8.9, Example 4 Use a graphing calculator to graph the inequality $8x + 3y > 24$.

First we write the related equation, $8x + 3y = 24$, and solve it for y. We get $y = -\frac{8}{3}x + 8$. We will enter this as $y1$. Press ◊ Y= to go to the equation-editor screen. If there is currently an entry for $y1$, clear it. Also clear or deselect any other equations that are entered. Now enter $y_1 = -\frac{8}{3}x + 8$. Since the inequality states that $8x + 3y > 24$, or y is *greater than* $-\frac{8}{3}x + 8$, we want to shade the half-plane above the graph of y_1. To do this, use the cursor to highlight $y1$. Then press 2nd F6 to display the

Style menu. Choose item 7, Above, by pressing 7. (To shade below a line we would press 8 to select Below.) Then press F2 6 to see the graph of the inequality in the standard viewing window.

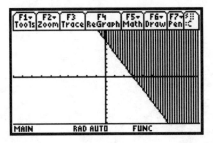

Note that when the "shade above" Style is selected it is not also possible to select the "Dot" style so we must keep in mind the fact that the line $y = -\frac{8}{3}x + 8$ is not included in the graph of the inequality. If you graphed this inequality by hand, you would draw a dashed line.

SYSTEMS OF LINEAR INEQUALITIES

We can graph systems of inequalities by shading the solution set of each inequality in the system with a different pattern. When the "shade above" or "shade below" style options are selected the calculator rotates through four shading patterns. Vertical lines shade the first function, horizontal lines the second, negatively sloping diagonal lines the third, and positively sloping diagonal lines the fourth. These patterns repeat if more than four functions are graphed.

Section 8.9, Example 8 Graph the system
$$x + y \leq 4,$$
$$x - y < 4.$$

First graph the equation $x + y = 4$, entering it in the form $y = -x + 4$. We determine that the solution set of $x + y \leq 4$ consists of all points on or below the line $x + y = 4$, or $y = -x + 4$, so we select the "shade below" style for this function. Next graph $x - y = 4$, entering it in the form $y = x - 4$. The solution set of $x - y < 4$ is all points above the line $x - y = 4$, or $y = x - 4$, so for this function we choose the "shade above" style. (See Example 4 above for instructions on selecting styles.) Now press F2 6 to display the solution sets of each inequality in the system and the region where they overlap in the standard viewing window. The region of overlap is the solution set of the system of inequalities. Keep in mine that the line $x + y = 4$, or $y = -x + 4$, is part of the solution set while $x - y = 4$, or $y = x - 4$, is not.

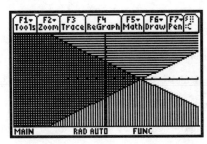

Chapter 9
Exponents and Radical Functions

RADICAL EXPRESSIONS AND RATIONAL EXPONENTS

As discussed in Section 9.2, we can enter an expression containing a square root using radical notation or rational exponents. For example, we can enter $y = \sqrt{x-3}$ using radical notation or as $y = (x-2)^{1/2}$ or as $y = (x-3)^{.5}$. To enter $= \sqrt{x-3}$, press $\boxed{\text{2nd}}$ $\boxed{\sqrt{}}$ $\boxed{\text{X}}$ $\boxed{-}$ 3 $\boxed{)}$. ($\sqrt{}$ is the second operation associated with the $\boxed{\times}$ multiplication key.) Note that the calculator supplies the left parenthesis along with the radical symbol and we add a right parenthesis after entering the radicand. To enter $y = (x-3)^{1/2}$, press $\boxed{(}$ $\boxed{\text{X}}$ $\boxed{-}$ 3 $\boxed{)}$ $\boxed{\wedge}$ $\boxed{(}$ 1 $\boxed{\div}$ 2 $\boxed{)}$. Note that both the radicand and the rational exponent are enclosed in parentheses. To enter $y = (x-3)^{.5}$, press $\boxed{(}$ $\boxed{\text{X}}$ $\boxed{-}$ 3 $\boxed{)}$ $\boxed{\wedge}$ $\boxed{.}$ 5. When the exponent is in decimal notation it is not necessary to enclose it in parentheses.

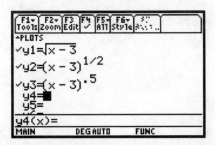

We use a rational exponent to enter a cube root as well as any other higher order roots. For example, to enter $y = \sqrt[3]{x+5}$ we enter $y = (x+5)^{1/3}$ by pressing $\boxed{(}$ $\boxed{\text{X}}$ $\boxed{+}$ 5 $\boxed{)}$ $\boxed{\wedge}$ $\boxed{(}$ 1 $\boxed{\div}$ 3 $\boxed{)}$. Since we cannot enter exact decimal notation for 1/3, we cannot use decimal notation for the exponent in this case.

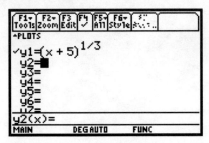

We can enter $f(x) = \sqrt[4]{2x-7}$, as in Section 9.2, Example 3, as $f(x) = (2x-7)^{1/4}$ or as $f(x) = (2x-7)^{.25}$.

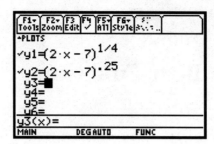

Chapter 10
Quadratic Functions and Equations

MAXIMUMS AND MINIMUMS; FINDING THE VERTEX

We can use a graphing calculator to find the vertex of a quadratic function. We do this by using the Maximum or Minimum feature from the Math menu or the Graph screen.

Section 10.7, Example 4 Use a graphing calculator to determine the vertex of the graph of the function given by $f(x) = -2x^2 + 10x - 7$.

The coefficient of x^2 is negative, so we know that the graph of the function opens down and, thus, has a maximum value. Clear or deselect any functions previously entered on the equation-editor screen. Then enter $y = -2x^2 + 10x - 7$. Choose a viewing window that shows the vertex. One good choice is $[-3, 7, -10, 10]$.

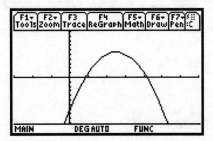

Now select the Maximum feature from the CALC menu by pressing $\boxed{F5}$ 4. We are prompted to select a lower bound for the vertex. Use the arrow keys to move the cursor to a point on the parabola to the left of the vertex or key in an x-value that is less than the x-coordinate of the vertex.

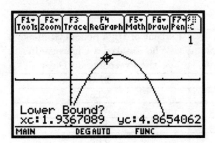

Press $\boxed{\text{ENTER}}$. Next we are prompted to select an upper bound. Move the cursor to a point on the parabola to the right of the vertex or key in an x-value that is greater than the x-coordinate of the vertex.

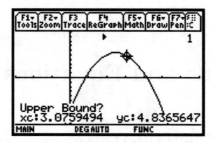

Press ENTER. We see that the maximum function value is 5.5, and it occurs when x is 2.5. Thus, the vertex of the graph of $f(x) = -2x^2 + 10x - 7$ is $(2.5, 5.5)$.

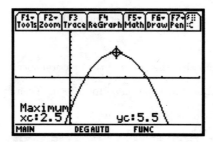

Minimum function values are found in a similar manner. Select the Minimum feature from the Math menu on the Graph screen by pressing F5 3.

QUADRATIC REGRESSION

Regression can be used to fit a quadratic function to data when three or more data points are given.

Section 10.8, Example 4(c) According to the Centers for Disease Control and Prevention, the percent of high school students who reported having smoked a cigarette in the preceding 30 days declined from 1997 to 2001, after rising in the first part of the 1990s. Use the REGRESSION feature of a graphing calculator to fit a quadratic function $H(x)$ to all the given data in the following table.

Years after 1991	Percent of High School Students Who Smoked a Cigarette in the Preceding 30 Days
0	27.5
2	30.5
4	34.9
6	36.4
8	34.9
10	28.5

We enter the data in the Data/Matrix editor as described on page 136 of this manual.

Then select QuadReg as the CalculationType from the Calc menu by pressing [F5] [▷] 9. Specify the sources of x and y and a function name to which the equation will be stored. (See page 137 of this manual for the procedure.) Then press [ENTER] [ENTER]. The calculator returns the coefficients of a quadratic function $y = ax^2 + bx + c$. We have $H(x) = -0.315179x^2 + 3.433214x + 26.507143$.

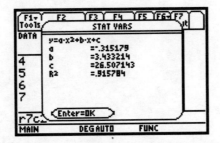

The function can be evaluated using one of the methods on pages 133 and 134.

Chapter 11
Exponential and Logarithmic Functions

COMPOSITE FUNCTIONS

For functions y_1 and y_2, when we enter $y_1(y_2)$ on a calculator we are entering the composition $y_1 \circ y_2$. The composite functions found in Section 11.1, Example 2 are checked using tables on a graphing calculator. To check that $f \circ g = \sqrt{x-1}$ when $f(x) = \sqrt{x}$ and $g(x) = x - 1$, enter $y_1 = \sqrt{x}$, $y_2 = x - 1$, $y_3 = \sqrt{x-1}$, and $y_4 = y_1(y_2)$ on the equation-editor screen. To enter y_4, position the cursor beside $y4 =$, clear any existing entry, and press $\boxed{Y}$ $\boxed{1}$ $\boxed{(}$ $\boxed{Y}$ $\boxed{2}$ $\boxed{(}$ $\boxed{X}$ $\boxed{)}$ $\boxed{)}$ $\boxed{\text{ENTER}}$. Then compare the values of y_3 and y_4 in a table. We show a table with tblStart = 1, Δtbl = 0.5, and Independent set on Auto. Use the $\boxed{\triangleright}$ key to scroll across the table to see the $y3$- and $y4$-columns.

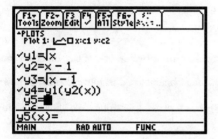

Similarly, to check that $g \circ f(x) = \sqrt{x} - 1$, also enter $y_5 = \sqrt{x} - 1$ and $y_6 = y_2(y_1)$. To enter y_6, position the cursor beside $y6 =$, clear any existing entry, and press $\boxed{Y}$ $\boxed{2}$ $\boxed{(}$ $\boxed{Y}$ $\boxed{1}$ $\boxed{(}$ $\boxed{X}$ $\boxed{)}$ $\boxed{)}$ $\boxed{\text{ENTER}}$.

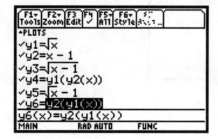

GRAPHING FUNCTIONS AND THEIR INVERSES

We can graph the inverse of a function using the DrawInv feature from the Draw menu on the Graph screen.

Section 11.1, Example 9(c) Graph the inverse of the function $g(x) = x^3 + 2$.

We will graph $g(x)$, $g^{-1}(x)$, and the line $y = x$ on the same screen. Press $\boxed{\diamond}$ $\boxed{Y=}$ to go to the equation-editor screen and clear or deselect any existing entries. Then enter $y_1 = x^3 + 2$ and $y_2 = x$. Select a square window by pressing $\boxed{F2}$ 5. Now paste the DrawInv command from the Draw menu to the home screen by pressing $\boxed{\text{2nd}}$ $\boxed{F6}$ 3. Indicate that we want to draw the inverse of y_1 by pressing $\boxed{Y}$ 1 $\boxed{(}$ $\boxed{X}$ $\boxed{)}$. Finally press $\boxed{\text{ENTER}}$ to see the graph of y_1^{-1} along with the graphs of y_1 and y_2. We show a window that has been squared from the standard window.

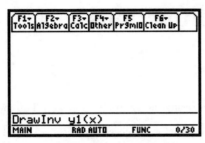
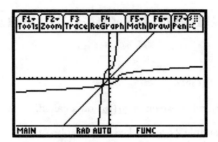

The drawing of y_1^{-1} can be cleared from the Graph screen by pressing $\boxed{\text{F4}}$ (ReGraph) or by pressing $\boxed{\text{2nd}}$ $\boxed{\text{F6}}$ to display the Draw menu and then pressing 1 to select ClrDraw. The ClrDraw command can also be accessed from the Catalog. From the home screen, press $\boxed{\text{CATALOG}}$ $\boxed{\text{C}}$, scroll to ClrDraw, and press $\boxed{\text{ENTER}}$ $\boxed{\text{ENTER}}$.

GRAPHING LOGARITHMIC FUNCTIONS

Section 11.3, Example 4 Graph: $f(x) = \log \dfrac{x}{5} + 1$.

We enter $y = \log(x/5) + 1$ on the equation-editor screen by positioning the cursor beside one of the function names and pressing $\boxed{\text{2nd}}$ $\boxed{\text{a-lock}}$ $\boxed{\text{L}}$ $\boxed{\text{O}}$ $\boxed{\text{G}}$ $\boxed{(}$ $\boxed{\text{X}}$ $\boxed{\div}$ 5 $\boxed{)}$ $\boxed{+}$ 1 $\boxed{\text{ENTER}}$. Note that $x/5$ must be enclosed in parentheses as shown. If parentheses are not used, the function entered will be $y = \dfrac{\log x}{5} + 1$. Clear or deselect any other functions. We show the function graphed in the window $[-2, 10, -5, 5]$.

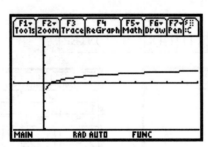

MORE ON GRAPHING

Section 11.5, Example 4 Graph: $f(x) = e^{-0.5x} + 1$.

We enter $y = e^{-0.5x}$ on the equation-editor screen by positioning the cursor beside one of the function names and pressing $\boxed{\diamond}$ $\boxed{e^x}$ $\boxed{(-)}$ $\boxed{.}$ 5 $\boxed{X}$ $\boxed{)}$ $\boxed{+}$ 1 $\boxed{\text{ENTER}}$. (Clear or deselect any other functions.) Select a window and press $\boxed{\text{GRAPH}}$. We show the function graphed in the window $[-5, 5, -2, 10]$.

TI-89

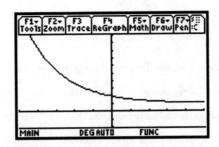

Section 11.5, Example 5(b) Graph: $f(x) = \ln(x + 3)$.

We enter $y = \ln(x + 3)$ on the equation-editor screen by positioning the cursor beside one of the function names and pressing 2nd LN X + 3) ENTER. (Clear or deselect any other functions.) Select a window and press ◊ GRAPH. We show the function graphed in the window $[-5, 10, -5, 5]$.

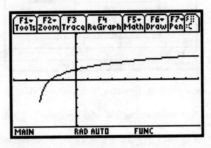

Section 11.5, Example 6 Graph: $f(x) = \log_7 x + 2$.

To use a graphing calculator we must first change the logarithmic base to e or 10. We will use e here. Recall that the change of base formula is $\log_b M = \dfrac{\log_a M}{\log_a b}$, where a and b are any logarithmic bases and M is any positive number. Let $a = e$, $b = 7$, and $M = x$ and substitute in the change-of-base formula. After clearing or deselecting previously entered functions, enter $y_1 = \dfrac{\ln x}{\ln 7} + 2$ on the equation-editor screen by positioning the cursor beside $y1 =$ and pressing 2nd LN X) ÷ 2nd LN 7) + 2 ENTER. Note that the parentheses must be closed in both the numerator and the denominator. Select a viewing window and press ◊ GRAPH. We show the graph in the window $[-2, 8, -2, 5]$.

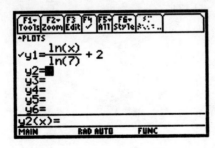 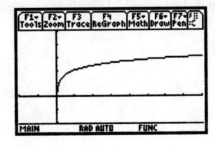

EXPONENTIAL REGRESSION

The TI-89 has an exponential regression feature.

Section 11.7, Example 9(a) In 1800, over 500,000 Tule elk inhabited the state of California. By the late 1800s, after the California Gold Rush, there were fewer than 50 elk remaining in the state. In 1978, wildlife biologists introduced a herd of 10 Tule elk into the Point Reyes National Seashore near San Francisco. By 1982, the herd had grown to 24 elk. There were 70 elk in 1986, 200 in 1996, and 500 in 2002. Use regression to fit an exponential function to the data and graph the function.

We enter the data as described on page 136 of this manual. Let x represent the number of years since 1978.

Now press $\boxed{F5}$ to go to the Calculate screen. Select ExpReg from the CalculationType menu by pressing $\boxed{\triangleright}$ 4. Enter the sources of x and y and the function name to which the equation will be saved. Then press $\boxed{\text{ENTER}}$. The calculator returns the values of a and b for the exponential function $y = ab^x$. We have $y = 13.016081(1.168548)^x$. We graph the equation in the window $[-2, 40, -5, 1000]$, xscl = 5, yscl = 100.

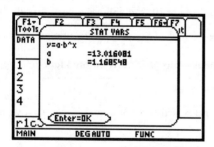 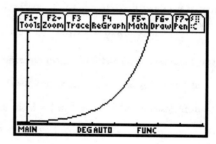

This function can be evaluated using one of the methods on pages 133 and 134.

Chapter 12
Conic Sections

GRAPHING CIRCLES

Because the TI-89 can graph only functions, the equation of a circle must be solved for y before it can be entered in the calculator. Consider the circle $(x-3)^2 + (y+1)^2 = 16$ discussed in Section 12.1. In the text it is shown that this is equivalent to $y = -1 \pm \sqrt{16 - (x-3)^2}$. One way to graph this circle is first to enter $y_1 = -1 + \sqrt{16 - (x-3)^2}$ and $y_2 = -1 - \sqrt{16 - (x-3)^2}$. Then select a square window, to eliminate distortion, and press $\boxed{\diamond}$ $\boxed{\text{GRAPH}}$. (See page 135 of this manual for a discussion on squaring the viewing window.) We show the graph in the window $[-7, 13, -6, 4]$.

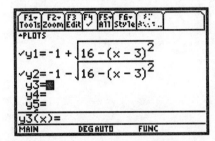
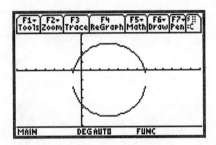

If the center and radius of a circle are known, the circle can be graphed using the Circle feature from the DRAW menu. Consider the circle $(x-3)^2 + (y+1)^2 = 16$ again. The center of this circle is $(3, -1)$ and its radius is 4. To graph it first press $\boxed{\diamond}$ $\boxed{\text{Y}=}$ and clear or deselect all previously entered equations on the equation-editor screen. Then select a square window. We will use $[-7, 13, -6, 4]$ as above. Now, to select Circle from the Catalog, first press $\boxed{\text{HOME}}$ or $\boxed{\text{2nd}}$ $\boxed{\text{QUIT}}$ to go to the home screen. Then press $\boxed{\text{CATALOG}}$ $\boxed{\text{C}}$ to go to the beginning of the items in the Catalog that start with C. (Note that it is not necessary to press $\boxed{\text{alpha}}$ before a letter key when the Catalog is displayed.) Now use $\boxed{\triangledown}$ to go to "Circle" and press $\boxed{\text{ENTER}}$. The Circle command appears on the entry line of the home screen. "Circle" can also be typed directly on the entry line of the home screen by pressing $\boxed{\text{2nd}}$ $\boxed{\text{a-lock}}$ $\boxed{\text{C}}$ $\boxed{\text{I}}$ $\boxed{\text{R}}$ $\boxed{\text{C}}$ $\boxed{\text{L}}$ $\boxed{\text{E}}$ $\boxed{\text{alpha}}$. Enter the x-coordinate of the center, the y-coordinate of the center, and the length of the radius, all separated by commas. To do this press 3 $\boxed{,}$ $\boxed{(-)}$ 1 $\boxed{,}$ 4. Press $\boxed{\text{ENTER}}$ to see the graph.

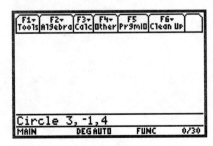
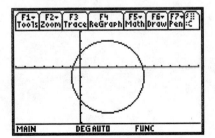

This graph can be cleared from the Graph screen by pressing $\boxed{\text{F4}}$ (ReGraph) or by pressing $\boxed{\text{2nd}}$ $\boxed{\text{F6}}$ to display the Draw menu and then pressing 1 to select ClrDraw. The ClrDraw command can also be accessed from the Catalog. From the home screen, press $\boxed{\text{CATALOG}}$ $\boxed{\text{C}}$, scroll to ClrDraw, and press $\boxed{\text{ENTER}}$ $\boxed{\text{ENTER}}$.

Chapter 13
Sequences, Series, and Probability

SEQUENCE MODE

To enter a sequence on a TI-89, first select Seq (Sequence) mode for the Graph setting. Press $\boxed{\text{MODE}}$ $\boxed{\triangleright}$ 4 $\boxed{\text{ENTER}}$.

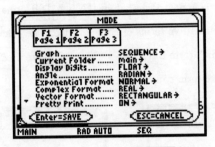

The function names that appear on the equation-editor screen when $\boxed{\diamond}$ $\boxed{Y=}$ is pressed are $u1$, $u2$, and so on rather than y_1, y_2, and so on. In addition we will use the variable n instead of x.

Section 13.1, Example 1 Find the first four terms and the 13th term of the sequence for which the general term is given by $a_n = (-1)^n n^2$.

After selecting Sequence mode, press $\boxed{\diamond}$ $\boxed{Y=}$ to go to the sequence-editor screen, enter the general term of the sequence beside "$u1 =$" by pressing $\boxed{(}$ $\boxed{(-)}$ 1 $\boxed{)}$ $\boxed{\wedge}$ $\boxed{\text{alpha}}$ $\boxed{\text{N}}$ $\boxed{\times}$ $\boxed{\text{alpha}}$ $\boxed{\text{N}}$ $\boxed{\wedge}$ 2 $\boxed{\text{ENTER}}$.

Now set up a table with Independent set to Ask. (See page 125 of this manual.) To see the first four terms and the 13th term of the sequence enter 1, 2, 3, 4, and 13 for n in the table.

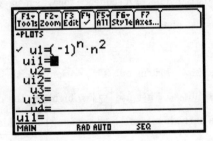

THE SEQUENCE FEATURE

The Sequence feature of the TI-89 writes the terms of a sequence as a list. This feature can be used even if the calculator is not in Sequence mode.

Section 13.1, Example 2 Use a graphing calculator to find the first five terms of the sequence for which the general term is given by $a_n = n/(n+1)^2$.

From the home screen we will access the Sequence feature from the Math List menu and copy it to the entry line by pressing [2nd] [Math] 3 1. Now enter the general term of the sequence, the variable, and the values of the variable for the first and last terms we wish to calculate, all separated by commas. Press [alpha] [N] [÷] [(] [alpha] [N] [+] 1 [)] [^] 2 [,] [alpha] [N] [,] 1 [,] 5 [)]. Now press [ENTER] to see a list of the first five terms of the sequence. The calculator is set in Auto mode, so the terms are expressed as fractions. In order to see the fifth term in the list we must first press [△] to move to the history area of the screen. Then we press [▷] repeatedly until the entire term can be seen.

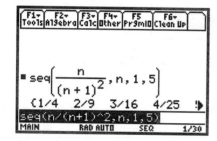 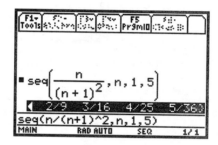

FINDING PARTIAL SUMS

We can use a graphing calculator to find partial sums of a sequence for which the general term is given by a formula.

Section 13.1, Example 5 Use a graphing calculator to find S_1, S_2, S_3, and S_4 for the sequence in which the general term is given by $a_n = (-1)^n/(n+1)$.

We will use the cumSum feature from the Math List menu. This option lists the cumulative, or partial, sums for a sequence defined using the Sequence feature discussed above. First use [▽] to highlight the current entry on the entry line of the home screen. Then copy cumSum to the entry line of the home screen by pressing [2nd] [Math] 3 7. Next copy the Sequence feature by pressing [2nd] [Math] 3 1. Now enter the general term of the sequence, the variable, and the first and last partial sums we wish to calculate, all separated by commas. Press [(] [(-)] 1 [)] [^] [alpha] [N] [÷] [(] [alpha] [N] [+] 1 [)] [,] [alpha] [N] [,] 1 [,] 4 [)] [)] [ENTER]. Note that we must use [△] to move to the history area of the screen and then press [▷] repeatedly until the entire fourth sum can be seen.

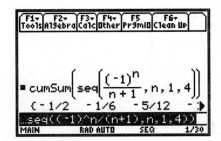 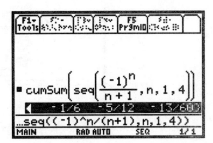

GRAPHS OF SEQUENCES

Section 13.1, Example 8 Graph the sequence for which the general term is given by $a_n = (-1)^n/n$.

The calculator must be set in Sequence mode to graph a sequence.

Press $\diamond$ $\boxed{Y=}$ to go to the sequence-editor screen, and enter $u1 = (-1)^n/n$ by positioning the cursor beside "$u1 =$" and pressing $\boxed{(}$ $\boxed{(-)}$ $\boxed{1}$ $\boxed{)}$ $\boxed{\wedge}$ $\boxed{\text{alpha}}$ $\boxed{N}$ $\boxed{\div}$ $\boxed{\text{alpha}}$ $\boxed{N}$.

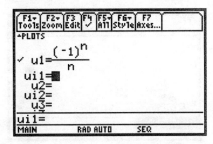

The domain of a sequence is a set of integers, so the graph of a sequence is a set of points that are not connected. Thus, we use Dot style to graph a sequence. Press $\boxed{\text{2nd}}$ $\boxed{\text{F6}}$ 2 to select "Dot" from the Style menu on the sequence-editor screen.

Next we enter the window dimensions. We will graph the sequence from $n = 1$ through $n = 15$, so we let nmin = 1, nmax = 15, xmin = 1, and xmax = 20. A table of values of the sequence shows that the terms appear to be between -1 and 1, so we let ymin = -1 and ymax = 1 with yscl = 0.1. We also set both plotStrt and plotStep to 1. These settings cause the graph to begin with the first term in the sequence and to plot each term of the sequence.

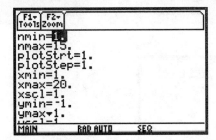 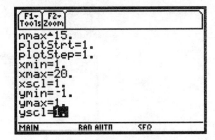

Press $\diamond$ $\boxed{\text{GRAPH}}$ to see the graph of the sequence.

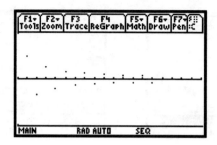

EVALUATING FACTORIALS

Factorials can be evaluated on a graphing calculator.

Section 13.4, Example 3 Simplify: $\dfrac{8!}{5!3!}$.

We use the factorial feature, denoted !, from the Math Probability menu. On the entry line of the home screen press 8 [2nd] [MATH] 7 1 [÷] [(] 5 [2nd] [MATH] 7 1 3 [2nd] [MATH] 7 1 [)] [ENTER]. Note that we must use parentheses in the denominator so that 8! is divided by both 5! and 3!.

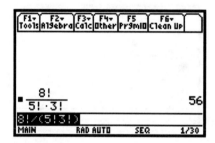

SIMPLIFYING $\binom{n}{r}$ NOTATION

Section 13.4, Example 4(a) Simplify: $\binom{7}{2}$.

The calculator uses the notation $_nC_r$ instead of $\binom{n}{r}$. To calculate $\binom{7}{2}$ on a TI-89 we use the $_nC_r$ feature from the Probability submenu of the Math menu. On the TI-89, $_7C_2$ is entered as $_nC_r(7,2)$.

Press [2nd] [MATH] 7 3 7 [,] 2 [)] [ENTER]. The first four keystrokes display the Math Probability menu and select item 3, $_nC_r$, from that menu. The remaining keystrokes enter the values for n and r separated by a comma, close the expression with a right parenthesis, and finally cause $_7C_2$ to be evaluated. The result is 21.

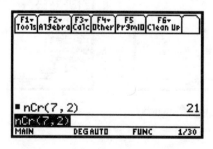

Index
TI-83 and TI-83 Plus Graphics Calculators

Absolute value, 6
Ask mode for a table, 9
Auto mode for a table, 10

Base, change of, 51
Binomial coefficients, 58

Catalog, 6, 23
Change of base formula, 51
Checking operations on polynomials, 27 – 29
Circles, graphing, 53
Clearing a drawing, 50, 53
Coefficients, binomial, 58
Combinations, 58
Composition of functions, 49
Contrast, 3
Copying regression equation to equation-editor screen, 23, 24
Converting to a fraction, 5
Cube root, 43
Cubic regression, 31
cumSum feature, 56
Curve fitting
    cubic regression, 31
    exponential regression, 52
    linear regression, 22
    quadratic regression, 46

Data, entering, 22
Deselecting an equation, 22
Determinants, evaluating, 39

Diagnostic On/Off, 23
Dot mode, 33
Dot graph style, 33
DRAW menu, 49, 53
DrawInv, 49

e^x, 50
Editing entries, 4
Entering data, 22
Entering equations, 9, 11
Equations
    deselecting, 22
    entering, 9, 11
    graphing, 14
    rational, 43
    solving, 15, 16
Evaluating determinants, 39
Evaluating expressions, 3, 29
Evaluating formulas, 9
Evaluating functions, 18
Exponential functions, 50
Exponential regression, 52
Exponents, 7, 43

Factorials, 58
Formulas, evaluating, 9
Frac operation, 5
Fraction, converting to, 5
Function values, 18
    maximum, 45

TI-83 and TI-83 Plus Index

minimum, 46
Functions
    composition, 49
    evaluating, 18
    exponential, 50
    graphing, 33, 49 - 51
    inverse, 49
    logarithmic, 50, 51
    maximum value, 45
    minimum value, 46
    rational, 33
    zeros, 16

Graph styles
    dot, 33
    path, 27
    shade above, 40
    shade below, 40
Graphing circles, 53
Graphing equations, 14
    points of intersection, 15
Graphing exponential functions, 50
Graphing functions
    exponential, 50
    inverse, 49
    logarithmic, 50, 51
    rational, 33
Graphing inequalities, 40
Graphing an inverse function, 49
Graphing logarithmic functions, 50, 51
Graphing rational functions, 33
Graphing sequences, 57

Grouping symbols, 7

Home screen, 3
Horizontal split-screen, 28

Inequalities
    graphing, 40
    solutions of, 40
    systems of, 40
Intersect feature, 15
Intersection, points of, 15
Inverse function, graphing, 49

Linear regression, 22
LIST menu, 56
Ln x, 51
Log x, 50, 51
Logarithmic functions, 50

Matrices
    row-equivalent operations, 38
    and solving systems of equations, 38
Maximum function value, 45
Menu, using, 4
Minimum function value, 46
Mode settings, 3
    dot, 33
    sequence, 55
    sequential, 27
    split-screen, 28, 29

TI-83 and TI-83 Plus Index

Modeling
    cubic regression, 31
    exponential regression, 52
    linear regression, 22
    quadratic regression, 46

n choose r, 58
Negative numbers, 6

On/Off, 3
Operations on polynomials, checking, 27 - 29
Order of operations, 7

Partial sums, 56
Path graph style, 27
Plot, turning off, 9
Plotting points, 24
Points of intersection, 15
Points, plotting, 24
Polynomial, evaluating, 29
Polynomial regression, 31

Quadratic function, vertex, 45
Quadratic regression, 46

Radical notation, 43
Rational exponents, 43
Rational functions, graphing, 33
Recalling an entry, 4
Reduced row-echelon form, 38

Regression
    cubic, 31
    exponential, 52
    linear, 22
    polynomial, 31
    quadratic, 46
Regression equation, copying to equation-editor screen, 23, 24
Row-equivalent operations, 38

Scatter diagram, 24
Scientific notation, 30
Sequence feature, 56
Sequence mode, 55
Sequences
    graphing, 57
    partial sums, 56
    terms, 55, 56
Sequential mode, 27
Shading, 40
Solutions of inequalities, 40
Solving equations, 15, 16
Solving systems of equations, 35, 38
Split-screen mode, 28, 29
Square roots, 5
Square viewing window, 21
Standard viewing window, 13
STAT lists, 22
STAT PLOT, 24
    turning off, 9
Storing values, 29

TI-83 and TI-83 Plus Index

Systems of equations
    graphical solutions, 35
    matrix solution, 38
Systems of linear inequalities, 40

Table feature
    ask mode, 9
    auto mode, 10
Terms of a sequence, 55, 56
Trace, 25
Turning off plots, 9

Value feature, 18
Values of functions, 18
VARS, 18, 23
Vertex of a quadratic function, 45
Vertical split-screen, 29
Viewing window, 13
    square, 21
    standard, 13

Window, *see* Viewing window

x^{th} root, 43

Zero feature, 17
Zeros of functions, 16
Zsquare (ZOOM 5), 21
Zstandard (ZOOM 6), 14

Index
TI-86 Graphics Calculator

Absolute value, 66
Ask mode for a table, 69, 70
Auto mode for a table, 71

Base, change of, 109
Binomial coefficients, 115

Catalog, 67, 107, 111
Change of base formula, 109
Checking operations on polynomials, 87, 88
Circle feature, 111
Circles, graphing, 111
Clearing a drawing, 108, 112
Coefficients, binomial, 115
Combinations, 115
Composition of functions, 107
Contrast, 63
Copying regression equation to equation-editor screen, 83
Converting to a fraction, 65
cSum feature, 114
Cube root, 101
Cubic regression, 91
Curve fitting
    cubic regression, 91
    exponential regression, 110
    linear regression, 82
    quadratic regression, 104
Custom menu, 67

Data, entering, 82
Deselecting an equation, 82
Determinants, evaluating, 99
DrawDot format, 93
DrInv, 107

e^x, 108
Editing entries, 64
Entering data, 82
Entering equations, 69, 72
Equations
    deselecting, 82
    entering, 69, 72
    graphing, 74
    solving, 75, 77
EVAL feature, 79
Evaluating determinants, 99
Evaluating expressions, 63, 88
Evaluating formulas, 69
Evaluating functions, 79, 85
Exponential functions, 108
Exponential regression, 110
Exponents, 68, 101
Expressions, evaluating, 63, 88

Factorials, 115
Forecast feature, 85
Formats
    DrawDot, 93
    Sequential, 87
Formulas, evaluating, 69

TI-86 Index

Frac operation, 65
Fraction, converting to, 65
Function values, 79, 85
    maximum, 103
    minimum, 104
Functions
    composition, 107
    evaluating, 79
    exponential, 108
    graphing, 93, 107 - 109
    inverse, 107
    logarithmic, 108, 109
    maximum value, 103
    minimum value, 104
    zeros, 77

Graph styles
    dot, 93
    path, 87
    shade above, 99
    shade below, 100
Graphing circles, 111
Graphing equations, 74
    points of intersection, 75, 95
Graphing exponential functions, 108
Graphing functions
    exponential, 108
    inverse, 107
    logarithmic, 108, 109
    rational, 93
Graphing inequalities, 99
Graphing an inverse function, 107

Graphing logarithmic functions, 108, 109
Graphing rational functions, 93
Graphing sequences, 114
Grouping symbols, 68

Home screen, 63

Inequalities
    graphing, 99
    solutions of, 99
    systems of, 100
Intersect feature, 75
Intersection, points of, 75
Inverse function, graphing, 107

Linear regression, 82
LIST menu, 113
Ln x, 109
Log x, 108, 109
Logarithmic functions, 108, 109

Matrices
    row-equivalent operations, 98
    and solving systems of equations, 97, 98
Maximum function value, 103
Menu
    custom, 67
    using, 64
Minimum function value, 104
Mode settings, 63

TI-86 Index

Modeling
    cubic regression, 91
    exponential regression, 110
    linear regression, 82
    quadratic regression, 103

n choose *r*, 115
Negative numbers, 66

On/Off, 63
Operations on polynomials, checking, 87, 88
Order of operations, 68

Partial sums, 114
Path graph style, 87
Plot, turning off, 69
Plotting points, 84
Points of intersection, 75
Points, plotting, 84
Polynomial, evaluating, 88
Polynomial regression, 91

Quadratic function, vertex, 103
Quadratic regression, 104

Radical notation, 101
Rational exponents, 101
Rational functions, graphing, 93
Recalling an entry, 64
Reduced row-echelon form, 98

Regression
    cubic, 91
    exponential, 110
    linear, 82
    polynomial, 91
    quadratic, 104
Regression equation, copying to equation-editor screen, 83
Root feature, 77
Row-equivalent operations, 98

Scatter diagram, 84
Scientific notation, 89
Sequence feature, 113
Sequences
    graphing, 114
    partial sums, 114
    terms, 113, 114
Sequential format, 87
Shading, 99, 100
Solutions of inequalities, 99
Solving equations, 75, 77
Solving systems of equations, 95, 97
Split-screen mode, 88
Square roots, 66
Square viewing window, 81
Standard viewing window, 73
STAT lists, 82
STAT PLOT, 84
    turning off, 69
Storing values, 88

TI-86 Index

Systems of equations
    graphical solutions, 95
    matrix solution, 97, 98
Systems of linear inequalities, 100

Table feature
    ask mode, 69, 70
    auto mode, 71
Terms of a sequence, 113, 114
Trace, 85
Turning off plots, 69

Values of functions, 79, 85
Vertex of a quadratic function, 103
Viewing window, 73
    square, 81
    standard, 73

Window, *see* Viewing window

x^{th} root, 101

Zeros of functions, 77
ZSQR, 81
ZSTD, 74

Index
TI-89 Graphics Calculator

Absolute value, 123, 124
Approximate mode, 122
Ask mode for a table, 125
Auto mode for a table, 127
Auto setting, 122, 123

Base, change of, 163
Binomial coefficients, 170

Catalog, 124, 165
Change of base formula, 163
Checking operations on polynomials,
    141 – 143
Circles, graphing, 165
Clearing a drawing, 162, 166
Clearing the home screen, 121
Coefficients, binomial, 170
Combinations, 170
Composition of functions, 161
Contrast, 119
Copying regression equation to
    equation-editor screen, 137
Cubic regression, 145
cumSum feature, 168
Curve fitting
    cubic regression, 145
    exponential regression, 164
    linear regression, 136
    quadratic regression, 158

Data, entering, 136

Data/Matrix editor, 136
Defining a plot, 137
Deselecting an equation, 136
Deselecting a plot, 125
Determinants, evaluating, 153
Dot graph style, 147
Draw menu, 161, 165
DrawInv, 161

e^x, 162
Editing entries, 121
Entering data, 136
Entering equations, 125, 128
Entry line, 119
Equations
    deselecting, 136
    entering, 125, 128
    graphing, 130
    solving, 131, 132
Evaluating determinants, 153
Evaluating expressions, 120, 143
Evaluating formulas, 125
Evaluating functions, 133, 134
Exact mode, 122
Exponential functions, 162
Exponential regression, 164
Exponents, 124, 155
Expressions, evaluating, 120, 143

Factorials, 170
Fraction notation, 122

TI-89 Index

Function values, 133, 134
    maximum, 157
    minimum, 158
Functions
    composition, 161
    evaluating, 133, 134
    exponential, 162
    graphing, 147, 161 - 163
    inverse, 161
    logarithmic, 162, 163
    maximum value, 157
    minimum value, 158
    zeros, 132

Graph styles
    Above, 153, 154
    Below, 154
    Dot, 147
    Path, 141
Graphing circles, 165
Graphing equations, 130
    points of intersection, 131, 149
Graphing exponential functions, 162
Graphing functions
    exponential, 162
    inverse, 161
    logarithmic, 162, 163
    rational, 147
Graphing inequalities, 153, 154
Graphing an inverse function, 161
Graphing logarithmic functions, 162, 163

Graphing rational functions, 147
Graphing sequences, 169
Grouping symbols, 124

History area, 119
Home screen, 119
Horizontal split-screen, 142

Inequalities
    graphing, 153, 154
    solutions of, 153
    systems of, 154
Intersection feature, 131
Intersection, points of, 131, 149
Inverse function, graphing, 161

Left-right split-screen, 142
Linear regression, 136
Ln x, 163
Log x, 162, 163
Logarithmic functions, 162, 163

Math List menu, 168
Matrices
    row-equivalent operations, 152
    and solving systems of equations, 151
Maximum function value, 157
Menu, using, 121
Minimum function value, 158
Mode settings, 119, 120
    sequence, 167

TI-89 Index

split-screen, 142, 143
Modeling
    cubic regression, 145
    exponential regression, 164
    linear regression, 136
    quadratic regression, 158

n choose r, 170
Negative numbers, 124

On/Off, 119
Operations on polynomials, checking, 141 - 143
Order of operations, 124

Partial sums, 168
Path graph style, 141
Plot
    defining, 137
    deselecting/turning off, 125
Plotting points, 137
Points of intersection, 131, 149
Points, plotting, 137
Polynomial, evaluating, 143
Polynomial regression, 145

Quadratic function, vertex, 157
Quadratic regression, 158

Radical notation, 155
Rational exponents, 155
Rational functions, graphing, 147

Recalling an entry, 121
Reduced row-echelon form, 152
Regression
    cubic, 145
    exponential, 164
    linear, 136
    polynomial, 145
    quadratic, 158
Regression equation, copying to equation-editor screen, 137
Row-equivalent operations, 152

Scatter diagram, 138
Scientific notation, 144
Sequence feature, 168
Sequence mode, 167
Sequences
    graphing, 169
    partial sums, 168
    terms, 167, 168
Shading, 153, 154
Solutions of inequalities, 153
Solving equations, 131, 132
Solving systems of equations, 149, 151
Split-screen mode, 142, 143
Square roots, 123
Square viewing window, 135
Standard viewing window, 129, 130
Status line, 119
Storing values, 143
Systems of equations
    graphical solutions, 149

TI-89 Index

    matrix solution, 151
Systems of linear inequalities, 154

Table feature
    ask mode, 125
    auto mode, 127
Terms of a sequence, 167, 168
Toolbar, 119
Top-bottom split-screen, 142
Trace, 139
Turning off plots, 125

Value feature, 134
Values of functions, 133 134
Vertex of a quadratic function, 157
Vertical split-screen, 142
Viewing window, 129
    square, 135
    standard, 129, 130

Window, *see* Viewing window

Zero feature, 132
Zeros of functions, 132
ZoomSqr, 135
ZoomStd, 130